NO BULL ~~NOBLE~~ REVIEW™

MACROECONOMICS AND MICROECONOMICS

For use with the AP® Macroeconomics and AP® Microeconomics Exams

A no-nonsense approach to prepare for class and the big tests

by Craig Medico

About the Author:

Craig Medico is an AP Economics teacher at Paul D. Schreiber High School in Port Washington, New York, and the Macroeconomics instructor for the Junior Statesmen Summer School at Princeton University. He is the developer of several iPhone test prep apps from Study By App, LLC, including Economics AP, Economics AP Free, and Economics Flashcard Review. Mr. Medico was recently featured on the Economics Rockstar podcast and has contributed to WNYC Radio/Public Radio International's morning news show, The Takeaway. Visit Mr. Medico's homepage at MrMedico.info for No Bull Economics videos, podcasts, and other educational resources.

The No Bull Approach (2016 Edition)

No Bull Review…"because your review book, shouldn't need a review book!"

This No Bull Review™ book is the most concise and to the point review that you will find for the AP® Macroeconomics and AP® Microeconomics Exams. Our goal here is to give you everything you need to know for class and standardized testing. Sometimes review books can be full of material that you just don't need to know. Or, they give explanations that are just as long as the ones found in the textbooks. The No Bull approach is to cut through the fat, and give you what you want.

We, as authors of No Bull Review, are teachers. For years, we have been speaking to students to find out what you want in a review book. The answer? No Bull. You want the facts, clear and to the point. And…you want review questions. Lots of them.

Each No Bull Unit Review chapter contains a brief recap of the major concepts, including all of the necessary graphs and diagrams that you need to know, followed by a worksheet, practice questions, and solutions. The first six chapters cover Macroeconomics and the final five chapters recap the main ideas of Microeconomics. Please note that the first chapter on Basic Concepts is required for both, Macroeconomics and Microeconomics, exams.

At the end of the book you will find two No Bull Exams, one for Macroeconomics and one for Microeconomics. You will also find two intense review sheets. If you know all of the terms, graphs, and formulas on the *No Bull Review Sheets,* you should find success.

Remember, this is a no-nonsense study guide designed to maximize your study time. We hope you enjoy the No Bull approach. Thank you, and best of luck!

-No Bull Review

Table of Contents

Macroeconomics

Microeconomics

No Bull Practice Exams

No Bull Review Sheets

Look for the new
No Bull Economics
tip boxes in each
review chapter for
quick and easy
concepts that you
must know

NB1. Basic Concepts (Macro/Micro) – Review

Let's begin with some of the most basic concepts that are necessary in understanding **macroeconomics** ("the big picture") and **microeconomics** ("the individual details in the big picture"). First off, there is the economic problem of **scarcity**. In every economy, there are people with unlimited wants; however, the economic resources or **factors of production** are limited.

There are four types of **economic resources: land** consists of an economy's natural resources; **labor** consists of hourly wage earners and salaried employees, **capital** includes tools, machines, and factories; and the **entrepreneur** is the risk taker that combines land, labor, and capital with the hopes of making an economic profit.

Market economies (markets allocate resources), **traditional economies** (customs allocate resources), and **command economies** (central planners allocate resources) all face the burden of scarcity and must answer these three questions: What to produce? How to produce? For whom to produce?

In economics, an **opportunity cost** is the next best alternative. For example, your opportunity cost of studying economics at this very moment might be catching up on much needed sleep. We reject the benefits of the next best alternative to gain even greater benefits.

The **production possibilities curve**, also known as the production possibilities frontier, is a great way of illustrating opportunity costs within an economy.

Let's assume that we have an economy that can produce only two goods: beach balls and ice cream cones *(see Diagram 1)*. In this model, we also assume that economic resources and technology are fixed in the present.

If the economy shown below produces 5 beach balls then it can produce 100 ice cream cones. If it increases production to 6 beach balls, it must sacrifice 8 ice cream cones (100-92). If it increases production to 7 beach balls, then it must sacrifice 14 ice cream cones (92-78).

Diagram 1: PPC Increasing Costs

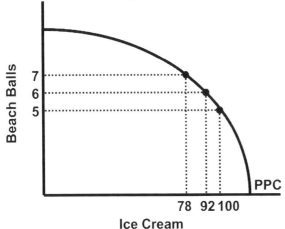

An economy's ***production possibilities curve*** illustrates ***scarcity*** and ***opportunity cost***

This is known as the **law of increasing opportunity costs**. To produce more beach balls, this economy must sacrifice increasing quantities of ice cream cones. The law of increasing costs is also reflected in the shape of the PPC, which is bowed outward from the origin. Here's another depiction of the production possibilities curve with increasing opportunity costs.

Diagram 2: E = full employment, U = unemployment, X = unattainable in the present

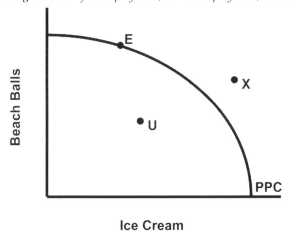

Resources are **specialized,** or not perfectly substitutable, when the **production possibilities curve** is bowed outward

Opportunity costs are increasing!

If the economy above is operating at Point E *(see Diagram 2)*, or anywhere along the curve, we can conclude that the economy is **fully employed** and achieving productive efficiency. **Productive efficiency** means producing at the lowest cost.

Point U, or anywhere inside the curve, indicates that there is **unemployment** or unused resources.

Point X, or anywhere outside the PPC, is impossible to attain in the present.

Let's shift our attention to the concept of **absolute advantage**. Say we have two economies: Surf Kingdom and Sand Land. Suppose that each economy can produce beach balls and ice cream cones with its economic resources.

Diagram 3: PPC Constant Cost

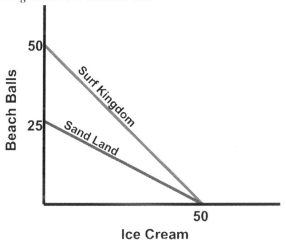

Resources are perfect substitutes when the **production possibilities curve** is a straight line

Opportunity costs are constant!

In this example *(see Diagram 3)*, the PPC is a straight line. This means that **opportunity costs are constant** and resources are perfectly substitutable in the production of these goods. If Surf Kingdom and Sand Land put all of their resources into ice cream production, they can both produce 50 ice cream cones. However, when Surf Kingdom puts all of its resources into beach ball production, it can produce 50 while Sand Land can only produce 25 *(see Diagram 4)*. Because Surf Kingdom can produce more beach balls than Sand Land, it has the absolute advantage in beach ball production. This also means that Surf Kingdom can produce one beach ball in less time than Sand Land and with fewer economic resources.

Diagram 4: Absolute Advantage Chart

	Beach Balls	Ice Cream Cones
Surf Kingdom	50	50
Sand Land	25	50

The economy that can produce more units of a good, one unit of a good faster, or one unit with fewer units of a *resource* has an *absolute advantage*

When it comes to **specialization and trade**, opportunity costs are most important. The opportunity cost of producing 1 beach ball in Surf Kingdom is 1 ice cream cone. That is because 50 ice cream cones divided by 50 beach balls is equal to 1. Sand Land's opportunity cost of producing one beach ball is 2 ice cream cones, or 50 ice cream cones divided by 25 beach balls. Because Surf Kingdom has a lower opportunity cost in beach ball production than Sand Land, it has a **comparative advantage**.

Diagram 5: Comparative Advantage Chart

	Beach Balls	Cost of 1 Beach Ball	Ice Cream Cones	Cost of 1 Ice Cream Cone
Surf Kingdom	50	1 Ice Cream Cone	50	1 Beach Ball
Sand Land	25	2 Ice Cream Cones	50	½ Beach Ball

The economy that can produce one unit with the lowest *opportunity cost* has a *comparative advantage*

As you can see in the chart above, Sand Land has a lower opportunity cost for ice cream cone production and therefore the comparative advantage in ice cream cones *(see Diagram 5)*.

If these two countries were to specialize and trade, Surf Kingdom would produce and export beach balls while importing ice cream cones from Sand Land; Sand Land would produce and export ice cream cones while importing beach balls from Surf Kingdom.

Trade will only happen if both countries gain from the exchange. For Surf Kingdom to gain from trade, it must receive more than 1 ice cream cone for each beach ball that it exports *(see Diagram 6)*.

Diagram 6: Developing Terms of Trade, Surf Kingdom

	Beach Balls	Cost of 1 Beach Ball	Ice Cream Cones	Cost of 1 Ice Cream Cone
Surf Kingdom	50	1 Ice Cream Cone	50	1 Beach Ball

Trade: Must receive more than 1 ice cream cone for each beach ball (or export less than 1 beach ball for each ice cream cone they import).

Opp. Cost of X (w/output):
$$\frac{\text{Total Units Y}}{\text{Total Units X}}$$
Opp. Cost of X (w/resources):
$$\frac{\text{Resource Units X}}{\text{Resource Units Y}}$$

Sand Land must receive more than ½ a beach ball for each ice cream cone it exports *(see Diagram 7)*.

Diagram 7: Developing Terms of Trade, Sand Land

	Beach Balls	Cost of 1 Beach Ball	Ice Cream Cones	Cost of 1 Ice Cream Cone
Sand Land	25	2 Ice Cream Cones	50	½ Beach Ball

Trade: Must receive more than ½ beach ball for each ice cream cone (or export fewer than 2 ice cream cones for each beach ball they import).

Terms of trade is beneficial for a country when:

$$\frac{\text{Imports}}{\text{Exports}} \; > \; \text{Opp. Cost of Export}$$

Therefore, acceptable terms of trade would be 1 beach ball for 1.5 ice cream cones. Sand Land receives 2/3 of a beach ball for each ice cream cone, while Surf Kingdom receives 1.5 ice cream cones for each beach ball. This would enable both economies to produce beyond their respective production possibilities curves.

Now it is time for the most recognizable words in the world of economics: **Supply and Demand**.

Suppose that the market for video games is currently in **equilibrium** *(see Diagram 8)*. This means that the demand curve intersects the supply curve at a specific price and quantity. The demand curve represents the buyers and the supply curve represents the sellers of video games. At high prices sellers want to sell more, and at low prices, buyers want to buy more. In this market, the equilibrium price of video games is $50 and the equilibrium quantity is 80,000 games. Please note that we will review these relationships with more detail in NB7.

Diagram 8: Market Equilibrium

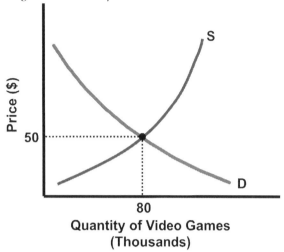

A change in price will cause point-to-point movement along the **supply** and **demand** curves

If the price of a game decreases to $20, there would be a temporary **shortage** *(see Diagram 9)*. The quantity demanded (QD) by consumers would exceed the quantity supplied (QS). In this example, there is a shortage of 70,000 games (QD of 105,000 minus QS of 35,000). Market forces will eventually push the price toward the equilibrium. It is important to note that a price change will only change the quantity supplied or quantity demanded; this is shown with point-to-point movement along the curves. **A price change in the current market will NOT shift the supply or demand curves.**

Diagram 9: Temporary Shortage

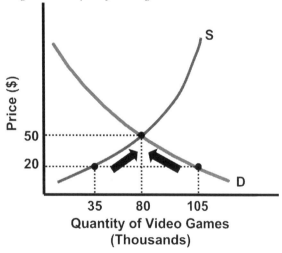

A **shortage** occurs when the QD>QS

QD temporarily moves point-to-point right and QS moves point-to-point left

If the price of a game increases to $99, there would be a temporary **surplus** *(see Diagram 10)*. The quantity supplied will exceed the quantity demanded. Now, there is a surplus of 70,000 video games (QS of 100,000 minus QD of 30,000). Market forces will eventually push the price down toward equilibrium.

Diagram 10: Temporary Surplus

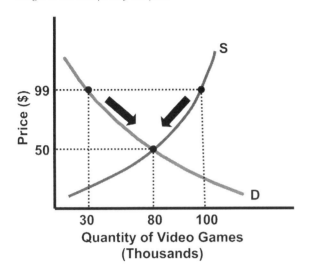

A ***surplus*** occurs when the QS>QD

QS temporarily moves point-to-point right and QD moves point-to-point left

The demand and supply curves can shift causing new market equilibriums to occur. The **determinants of demand** that shift the demand curve to the left or right include changes in: tastes and preferences; consumer income (more income increases demand for **normal goods**, but decreases demand for **inferior goods**); the number of buyers; expectations of future prices; prices of substitute goods; and prices of complementary goods.

Here is an example of an **increase in demand** for video games brought on by a decrease in the price of video game consoles, a **complementary good** to video games *(see Diagram 11)*. As a result of this rightward shift in demand, the market price increases to $60 and market quantity increases to 90,000 video games.

Diagram 11: Increase in Demand: Price Increases, Quantity Increases

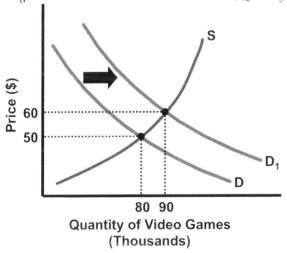

A change in price will never shift the ***demand curve***, but a shift of the demand curve will cause a change in price

Here is an example of a **decrease in demand** brought on by a decrease in the price of a

substitute good, say tablet computers *(see Diagram 12)*. When demand shifts left, the market price will decrease and the equilibrium quantity will also decrease.

Diagram 12: Decrease in Demand: Price Decreases, Quantity Decreases

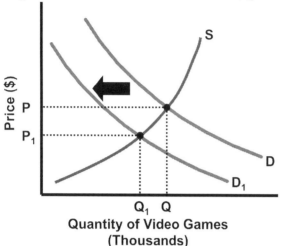

Price and quantity will increase when **demand shifts right**

Price and quantity will decrease when **demand shifts left**

The key **determinants of supply** include changes in: resource costs, per-unit production taxes and subsidies, number of sellers, productivity, technology, future price expectations, and the prices of other goods that use the same economic resources.

Here is an **increase in market supply** brought on by an increase in **productivity** *(see Diagram 13)*. As a result of the rightward shift in supply, the market price of a video game will decrease to $28 and the quantity will increase to 95,000 video games.

Diagram 13: Increase in Supply: Price Decreases, Quantity Increases

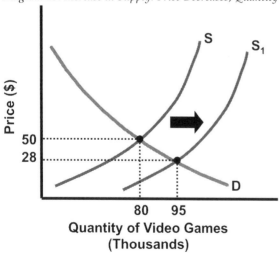

A change in price will never shift the **supply curve**, but a shift of the supply curve will cause a change in price

In the next example, **supply decreases** *(see Diagram 14)*. This leftward shift of supply could have been caused by higher factor costs or an increase in taxes on production. The market price will increase and the quantity will decrease.

Diagram 14: Decrease in Supply: Price Increases, Quantity Decreases

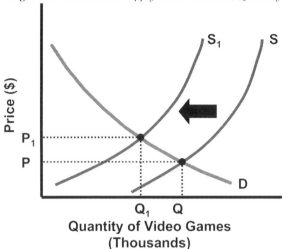

Price will decrease and quantity will increase when ***supply shifts right***

Price will increase and quantity will decrease when ***supply shifts left***

Here is an example containing **dual shifts**. Say the demand and supply curves shift right at the same time *(see Diagram 15)*. The market quantity will definitely increase, but what will happen to the price? If demand shifts right on its own, the market price increases. If supply shifts right on its own, the market price decreases. The market price can increase, decrease, or remain the same. Therefore, we say that price is **indeterminate**. When dual shifts occur, and you do not know the magnitude of the shifts, the price or quantity will be indeterminate *(see Diagram 15a)*.

Diagram 15: Demand Increases, Supply Increases: Price Indeterminate, Quantity Increases

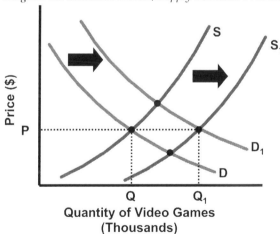

When the supply and demand curves shift in the same direction, the price will be ***indeterminate***

Diagram 15a: Dual Shifts Summary Chart

Shifts	Price	Quantity
Demand Increases, Supply Increases	Indeterminate	Increases
Demand Decreases, Supply Decreases	Indeterminate	Decreases
Demand increases, Supply decreases	Increases	Indeterminate
Demand decreases, Supply increases	Decreases	Indeterminate

When the supply and demand curves shift in opposite directions, the quantity will be *indeterminate*

Sometimes the government passes laws that prevent the market equilibrium from occurring. If the government wants to help low-income consumers, it might establish an effective **price ceiling**, a maximum legal price below the free-market equilibrium *(see Diagram 16)*. It makes video games more affordable for some buyers; however, this type of price control will lead to a shortage because the quantity demanded exceeds the quantity supplied. Sellers may choose to exit the industry in the long run, which will worsen this shortage. **Underground market** or illegal market activity may also become more rampant in this industry.

Diagram 16: Price Ceiling

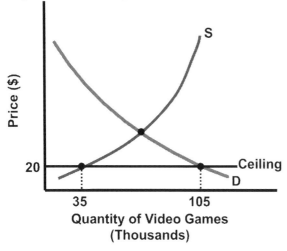

A *price ceiling* set below the equilibrium price causes a *shortage*

A price ceiling set above the equilibrium price does NOTHING

If the government wants to support low-income sellers, then it might establish a **price floor**, a minimum price above the equilibrium *(see Diagram 17)*. The problem with this price control is that it will lead to a market surplus. This is highly inefficient because it misallocates economic resources and thus causes deadweight loss.

Diagram 17: Price Floor

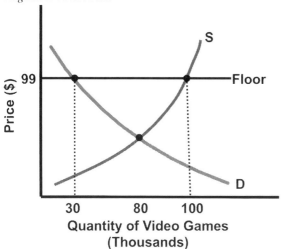

A ***price floor*** set above the equilibrium price causes a ***surplus***

A price floor set below the equilibrium price does NOTHING

These are the major concepts from the first unit of Macroeconomics and Microeconomics. You should test your skills by completing the worksheet and practice questions on Basic Concepts. Following the practice questions, you will find answers and explanations.

NB1. Basic Concepts (Macro/Micro) – Worksheet

Identify the following terms:

a. Scarcity _unlimited wants; limited resources_

b. Opportunity Cost _second best use of time_

c. Production Possibilities Curve _graph of opportunity cost_

d. Law of Increasing Opportunity Cost _concave PPC, costs more for next_

e. Demand _wants_

f. Supply _items_

g. Shortage _Demand > supply_

h. Surplus _Demand < supply_

i. Price Ceiling _artificial maximum price (shortage)_

j. Price Floor _artificial minimum price (surplus)_

Summary questions:

1. How do the studies of macroeconomics and microeconomics differ?

2. What are the four factors of production? _land, labor, capital, entrepreneurship_

3. How are market economies different from command economies?
   power in gov vs power in people

4. How do you illustrate a production possibilities frontier for an economy with constant opportunity costs?
   downward line

5. What is the difference between absolute advantage and comparative advantage?
   abs = whoever can produce more
   comp: lower opportunity cost

16

6. How do you determine acceptable trade terms using the law of comparative advantage?
   _____ Import/Export → opportunity cost of export _____
   _____

7. How is a change in quantity demanded different from a change in demand? _____
   _____ ∆ price _____ ∆ shift _____
   _____

8. What are the shift factors of demand? _____ future, changes in tastes, _____
   _____ consumer income, prices of supplements/complements, _____
   _____ goods. _____

9. How is a change in quantity supplied different from a change in supply? _____
   _____ ∆ price _____ ∆ shift _____
   _____

10. What are the shift factors of supply? _____ resource costs, govt tax/subsidy _____
    _____ technology, future expenses _____
    _____

NB1. Basic Concepts (Macro/Micro) – Practice

1. According to basic economic teachings, societies have limited resources and unlimited wants. This burden is known as
   a. scarcity
   b. ceteris paribus
   c. the fallacy of composition
   d. the crowding out effect
   e. unemployment

2. Which of the following is synonymous with Factors of Production?
   a. Goods and Services
   b. Output
   c. Economic Resources
   d. Supply
   e. Long-Run Aggregate Supply

3. Which of the following statements is true concerning command economies and mixed market economies?
   I. Economic decisions are primarily made by local customs within a command economy and markets help determine the prices of goods in a mixed market economy.
   II. The government solely determines what is produced in a command economy and tribal leaders allocate resources in a mixed market economy.
   III. The government and markets decide what is produced in a mixed market economy, while central planners determine how much to produce in a command economy.
   IV. Resources are scarce in both command and mixed market economies.
   a. I only
   b. II only
   c. II & IV
   d. III & IV
   e. IV only

4. Suppose Alex opens his own custom woodworking business. If Alex can build one jewelry box in 35 hours or two gun racks in 70 hours. What is his opportunity cost of producing one jewelry box?
   a. 0 gun racks
   b. 1/3 gun rack
   c. 1/2 gun rack
   d. 1 gun rack
   e. 2 gun racks

5. If the economy is currently operating at point B, what is the opportunity cost of producing one more guitar?

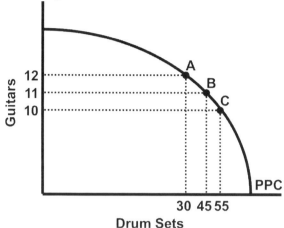

   a. 10 drum sets
   b. 15 drum sets
   c. 30 drum sets
   d. 45 drum sets
   e. 55 drum sets

6. Assume that Frankland can produce either 25 cars or 100 mopeds in one year, while Dragonia can produce 100 cars or 150 mopeds in one year. Which of the following is true?
   I. Frankland has the absolute advantage in car production.
   II. If the two countries were to specialize and trade, then Dragonia would import mopeds.
   III. Frankland has the comparative advantage in car production.
   IV. Dragonia has the absolute advantage in moped production, but not car production.
   a. II only
   b. II and III
   c. I and II
   d. III only
   e. I and IV

7. The graph below depicts the market for fluffy teddy bears. Which of the following is true at a price of $22?

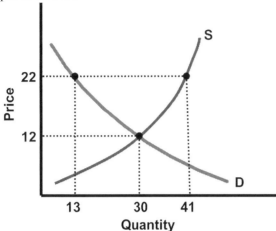

   a. There is a shortage of 10 fluffy teddy bears
   b. There is a shortage of 28 fluffy teddy bears
   c. There is a surplus of 28 fluffy teddy bears
   d. There is a shortage of 41 fluffy teddy bears
   e. There is a surplus of 41 fluffy teddy bears

8. If the supply of good H increases and the demand for good H increases simultaneously, then what will happen to market price and quantity?
   a. Price increases, Quantity increases
   b. Price increases; Quantity indeterminate
   c. Price indeterminate; Quantity decrease
   d. Price decrease; Quantity indeterminate
   e. Price indeterminate; Quantity increases

9. Refer to the diagram below. Assume that shoes and socks are complements and shoes are a normal good. Which of the following could have caused this demand curve to shift in the shoe market?

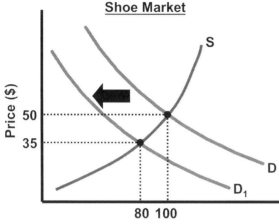

Shoe Market

I. The price of socks increased.
II. The price of sandals, a substitute good, increased.
III. Consumer incomes increased.
IV. The price of shoes decreased.
    a. I only
    b. II only
    c. I and II
    d. IV only
    e. III and IV

10. Suppose the government wants to help low income producers in the clothing market. The government decides that the best course of action is to establish a price floor that is below the free market price. What is the likely effect of this price control?
    a. A surplus of clothes
    b. A shortage of clothes
    c. Quantity supplied will increase
    d. Quantity supplied will decrease
    e. Nothing. Market will remain in equilibrium.

NB1. Basic Concepts (Macro/Micro) – Practice Answers

1. A, Scarcity. Scarcity is the economic problem that all societies face due to limited factors of production, and therefore must make economic decisions.

2. C, Economic Resources. Factors of production, economic resources, and economic inputs are different ways of referring to land, labor, capital, and entrepreneurship.

3. D, III & IV. Resources are scarce no matter the type of economy. In a command economy, central planners make decisions. In a mixed market system, markets and the government both allocate resources.

4. D, 1 gun rack. Since Alex can build one gun rack in 35 hours, his opportunity cost of producing one jewelry box is 1 gun rack (35/35).

5. B, 15 drum sets. If the economy is producing 11 guitars and wants to increase production to 12 guitars, then it must sacrifice 15 drums sets (45-30).

6. A, II only. Because Frankland has a lower opportunity cost than Dragonia in moped production (.25 car vs .67 car), Frankland would specialize in mopeds and export them to Dragonia.

7. C, There is a surplus of 28 fluffy teddy bears. Quantity supplied (41) minus quantity demanded (13) yields the surplus (28).

8. E, Price indeterminate; Quantity increases. If supply shifts right, price decreases and quantity increases. If demand shifts right, price increases and quantity increases. Price is indeterminate because it can increase, decrease, or remain constant. Quantity will definitely increase.

9. A, I only. If the price of a complementary good like socks rises, then demand for shoes will shift left.

10. E, Nothing. Market will remain in equilibrium. In order for a price floor to be effective, the legal minimum must be placed about the original equilibrium price. Here, the government placed the price floor below equilibrium so nothing will happen.

NB2. Economic Performance (Macro) – Review

Let's begin our review of Macroeconomics with a very important incentive system known as the **circular flow model** *(see Diagram 18)*. This model illustrates how households and businesses interact through the product market and factor market.

The upper half of the model shows households providing businesses with resources such as land, labor, capital, and entrepreneurship through the **factor market**. Households are willing to offer these resources because the businesses pay households in the forms of **income, interest, rent, and profit**.

The lower half of the Circular Flow Model depicts the movement of goods and services from businesses to households through the **product market**. The only reason businesses are willing to do this is because households pay businesses for goods and services in the form of **consumption**. This becomes **revenue** for the businesses. As you can see, money flows clockwise, while goods and economic resources flow counter clockwise in this depiction.

Diagram 18: Circular Flow Model

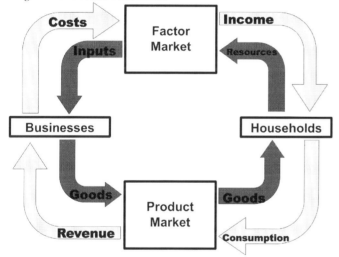

Costs are payments for economic resources

Consumer expenditures are payments for final goods and services

We can expand this model to include the **government** in the center, which buys resources from households and goods from businesses. The government provides goods and services to the households and businesses, and also collects taxes from both.

The **gross domestic product** (GDP) measures a nation's output in a given year by adding up the overall expenditures or incomes in an economy. It is a monetary value that only includes the production of final goods and services within a nation's borders.

Diagram 19: GDP Formula

$$GDP = C + I_g + G + X_n$$

The **expenditures formula** for a mixed market economy is Consumption + Gross Investment + Government Spending + Exports – Imports. Consumption is the largest component of GDP while Net Exports is the smallest *(see Diagram 19)*.

Consumption includes household spending while **gross investment** includes business expenditures on capital goods, residential and non-residential construction, intellectual property, and adjustments to inventories. Together, C and I_g make up the **private sector**.

GDP excludes financial transactions like stocks and bonds; transfer payments like social security checks; used goods like secondhand golf clubs; non-market work like you mowing your lawn; illegal work like bootlegging; unreported activity like the tips you forgot to report to the IRS; intermediate goods like the wood in your ukulele; and goods produced in other countries.

In order to determine if the economy is expanding, it is important to look at changes in **real GDP** over time. This measures the overall growth in output over time by deflating or inflating the **nominal GDP** (GDP that does not account for changes in the price level). This is done using a **GDP price index,** which is also known as the GDP deflator *(see Diagram 20)*.

Real GDP is equal to nominal GDP divided by the GDP price index in hundredths. If the price index is not given in hundredths, then we have to multiply by 100.

Diagram 20: Real GDP Formula

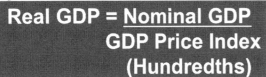

Real GDP = Nominal GDP
GDP Price Index
(Hundredths)

GDP does not account for income distribution, product quality, leisure time, underground markets, and pollution

The **business cycle** illustrates the general ups and downs of economic activity *(see Diagram 21)*. When real GDP is increasing and unemployment is falling, the economy is in the **expansionary phase**. Real GDP reaches its maximum at the **peak**.

Eventually real GDP and prices will fall. This occurs as the economy **contracts** and enters a recession. Inventories and unemployment will rise.

Diagram 21: Business Cycle

Peak

Peak

Expansion Contraction Expansion Expansion

Trough

Expansions: increasing real GDP, price level, and employment

Contractions: decreasing real GDP, price level, and employment

When real GDP bottoms out, the economy has reached the **trough**. The economy will eventually expand. Workers will be hired back, prices will rise, and real output will increase. The thin upward sloping line through the business cycle symbolizes the trend of economic growth.

By definition, an **unemployed person** is one that is over 16, non-institutionalized, and actively seeking employment. The **labor force** is comprised of those people looking for work and those that are currently working, including part-time and full-time workers.

To calculate the **unemployment rate**, take the number of unemployed people and divide by the labor force. To make it a percentage, simply multiply by 100 *(see Diagram 22)*. However, there is one big problem: the unemployment rate that you read about every month is understated because it does not include those people that have given up looking for work. They are known as **discouraged workers**. The official "U3" unemployment rate also counts part-time workers as fully employed.

Diagram 22: Unemployment Rate Formula

$$\text{Unemployment Rate} = \frac{\text{\# Unemployed}}{\text{\# Labor Force}} \times 100$$

> An ***unemployed person*** is one who looks for a job and isn't employed

There are three types of unemployment. One type is known as **frictional unemployment**, which consists of people between jobs or recent college grads looking for their first jobs. We can also include seasonal unemployment in this category because frictional unemployment is temporary and natural.

Structural unemployment is also natural. This exists when someone loses his or her job because the skills of the individual are no longer needed. These people will have to retrain or move to find new work. Structural and frictional unemployment make up the **natural rate of unemployment** at the full employment level of output.

The worst of the three types is **cyclical unemployment**. This is when people are laid off because the economy is in recession. As the unemployment rate increases beyond its natural rate, cyclical unemployment now exists and the economy is most likely in recession.

Inflation simply means that prices are rising. The effect of rising prices is that your money has less purchasing power. You can buy less stuff with the same income that you had in the past.

We can measure the **inflation rate** by examining changes in the **consumer price index,** which tracks a market basket of common consumer goods and services over time *(see Diagram 23)*. To get the percentage change in the CPI, subtract the previous CPI from the current CPI, divide by the previous CPI, and multiply by 100. In other words: "new minus old over old times 100."

Diagram 23: Inflation Rate Formula

$$\text{Inflation Rate} = \frac{\text{New CPI} - \text{Old CPI}}{\text{Old CPI}} \times 100$$

> ***Debtors*** can benefit from sudden inflation while ***lenders*** can lose

The two types of inflation are demand-pull and cost-push inflation. **Demand-pull inflation** exists when there is an increase in aggregate demand and prices rise. This can happen if the money supply grows faster than the production of goods and services.

The scarier type of inflation is known as **cost-push inflation**. This occurs when production costs increase or a **negative supply shock** occurs. Aggregate supply shifts left and we end up with higher prices, but, we also end up with more unemployment. This is a condition known as **stagflation**, which is a very difficult economic situation for policy makers to handle.

Inflation can help and hurt people. People that benefit from **unanticipated inflation** include those that make fixed payments. These debtors are essentially paying off loans with cheaper dollars. The real value of their debt is now worth less.

People that are harmed by unanticipated inflation include those that earn fixed incomes, and creditors who are receiving the cheaper dollars from the debtors.

Prices generally rise over time, but sometimes prices fall. **Deflation** occurs when the inflation rate becomes negative. Be careful not to confuse this concept with disinflation. **Disinflation** is when the rate of inflation slows down.

The easiest and perhaps most valuable equation from this unit is known as the **Fisher equation**. It defines the **real interest rate** and shows how inflation or expected inflation affects the real interest rate. Businesses that invest in capital goods and households that borrow and/or save money should pay close attention to this equation.

Diagram 24: Fisher Equation

Real I.R. = Nominal I.R. – Inflation Rate

Inflation can reduce *real interest rates* and *real wages*

The real interest rate is equal to the nominal interest rate minus the inflation rate *(see Diagram 24)*. You can also rewrite the equation to read: the nominal interest rate is equal to the real interest rate plus the inflation rate. This equation also applies to changes in real income. The percentage change in real income is equal to the percentage change in nominal income minus the inflation rate.

Those are the major concepts from unit two of Macroeconomics. Go test your skills by completing the worksheet and practice questions on Economic Performance. Following the practice questions, you will find answers and explanations.

NB2. Economic Performance (Macro) – Worksheet

Identify the following terms:

a. Circular Flow Model _model depicting money flow_

b. Gross Domestic Product _all goods i services prod in country/year_

c. Gross Investment _money spent by corporations for equipment_

d. Real GDP _GDP adjusted for inflation_

e. Business Cycle _troughs i peaks/ up lin full employment_

f. Frictional Unemployment _between jobs) college school going etc_

g. Structural Unemployment _people skill not obsolete new_

h. Cyclical Unemployment _depress i recess cycle low demand_

i. Inflation _rising rate cost push, demand pull_

j. Fisher Equation _Real IR . Nominal IR . inflation_

Summary questions:

1. How do the roles of households and businesses differ in the product market? _____
   households consume, business provide
   _____

2. How do the roles of households and businesses differ in the factor market? _____
   _bl-gm_
   _____

3. What are the components of gross domestic product according to the expenditures
   formula? _C I G+m_ _GDP: (+I+G+x-m_
   consumer investment gov't i export - imp
   _____

4. What is the difference between nominal GDP and real GDP? How do you calculate real
   GDP? _nominal dollar value_ _Real GDP. nominal_
   real - inflation cost
   _____

5. How do you determine the unemployment rate? ___ _# unemployed_ ___
   # labor force
   _____

6. What is the difference between an unemployed person and a discouraged worker? _____

discouraged doesn't _look_ _____

7. How do you determine the rate of inflation using the consumer price index? _____

CPI - 100 = inflation

New CPI - Current price - ...

base price

8. What is the difference between demand-pull inflation and cost-push inflation? _____

↑ demand pulls prices ↑

↑ cost pushes prices ↑

9. Which groups benefit from unanticipated inflation? _____

debtors

10. Which groups do not benefit from unanticipated inflation? _____

lenders, people with fixed income

NB2. Economic Performance (Macro) – Practice

1. Which flow best represents land, labor, and capital in the circular flow diagram shown below?

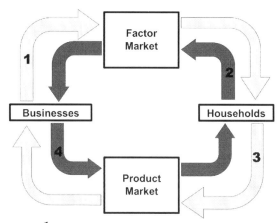

   a. 1
   b. 2
   c. 3
   d. 4
   e. All of the above

2. Which of the following is not included in calculating the gross domestic product of the United States?
   a. The salary of a US mailman
   b. An automobile produced in Texas
   c. An American motorcycle exported to Italy
   d. Monthly rent that is paid for an apartment built last year
   e. A TV built in China by an American company

3. A car produced by an American-owned factory in Taiwan is _____ in the calculation of Taiwan's GDP because _____.
   a. included; it is produced within Taiwan's borders
   b. included; it is a government expenditure
   c. not included; it is included in the USA's GDP
   d. not included; it is an import
   e. not included; it represents gross investment in the USA's GDP

4. Assume that the nominal GDP is $200 billion and the GDP price index is 0.5. What is the real GDP?
   a. $100 billion
   b. $200 billion
   c. $300 billion
   d. $400 billion
   e. $500 billion

5. Assume that the unemployment rate has declined from 11% to 7.5% over the last year, and the consumer price index has increased at a rate of 2.5% over the same time period. Which phase of the business cycle is this economy most likely facing?
   a. Expansion
   b. Trough
   c. Contraction
   d. Peak
   e. Stagflation

6. Suppose that there are 100 million people that are currently employed and 50 million people that are unemployed and searching for employment. If the total population is 300 million people, what is the current rate of unemployment?
   a. 2%
   b. 12%
   c. 29%
   d. 33%
   e. 50%

7. Which of the following will most likely cause demand-pull inflation when the economy is operating at the full-employment level of output?
   a. An increase in personal income taxes
   b. An increase in the price of oil
   c. An increase in consumption
   d. An increase in the prime interest rate
   e. An increase in imports

8. Which of the following will most likely cause cost-push inflation?
   a. An increase in the price of oil
   b. An increase in the international value of the dollar
   c. An increase in the federal funds rate
   d. An increase in gross investment
   e. An increase in the discount rate

9. Assume that the consumer price index increased from 100 to 150 between year one and year two. What is the rate of inflation for year two?
   a. 1.5%
   b. 25%
   c. 50%
   d. 67%
   e. 150%

10. Assume that the inflation rate is 7% and the real interest rate is 9%. What is the nominal interest rate?
    a. -2%
    b. 2%
    c. 9%
    d. 13%
    e. 16%

NB2. Economic Performance (Macro) – Practice Answers

1. B, 2. Land, labor, and capital are economic resources, which are sold to businesses in the factor (resource) market.

2. E, TV built in China by an American company. The GDP only includes goods produced within a nation's borders. You can either add up all expenditures or incomes to calculate the GDP. The rental payment for an apartment would count as a service that the landlord is providing.

3. A, included; it is produced within Taiwan's borders. What matters most in the calculation of GDP is where the goods are produced. The gross national product cared about ownership, but GDP cares about location.

4. D, $400 billion. Real GDP is equal to the nominal GDP divided by the GDP price index in hundredths ($200/0.5 = 400$).

5. A, Expansion. With the falling unemployment rate and a moderate rise in consumer inflation, it is safe to say this economy is in an expansionary phase. It is most likely not at the peak yet because unemployment is still above the natural rate (about 5%).

6. D, 33%. The unemployment rate equals the number unemployed divided by the number in the labor force, then multiply by 100 ($50/150 = 0.33$; $0.33 \times 100 = 33\%$). The labor force includes people looking for work.

7. C, An increase in consumption. An increase in consumer expenditures when the economy is at full employment will cause demand-pull inflation (aggregate demand shifted right).

8. A, An increase in the price of oil. An increase in the price of an economic resource like oil will cause cost-push inflation (short-run aggregate supply shifted left).

9. C, 50%. To calculate consumer inflation: (current CPI of 150 – original CPI of 100) divided by the original CPI of 100. Lastly, multiply the final answer by 100.

10. E, 16%. The nominal interest rate equals the real interest rate plus the inflation rate (9% + $7\% = 16\%$).

NB3. AD/AS & Fiscal Policy (Macro) – Review

In this section, we will review the key ideas of the **aggregate demand and aggregate supply** model, as well as fiscal policy.

Let's begin by graphing an economy operating at the full-employment level of output using a **three-segmented aggregate supply curve** *(see Diagram 25)*. In order to properly label the axes, make sure you keep the GDP *real* on the horizontal axis, and the price *level* on the vertical axis.

Diagram 25: Full-Employment Equilibrium with a 3-Segmented AS Curve

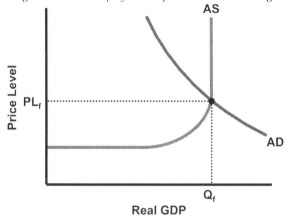

> The ***spending multiplier*** is most powerful when ***aggregate supply*** is horizontal

The horizontal range of the AS curve represents the **Keynesian range** and the upward sloping range represents the **intermediate range**. Together, these two ranges make up the **short-run aggregate supply curve**. The vertical range represents the **classical range** and indicates the full-employment level of output. To show full-employment, the aggregate demand curve intersects AS at the very bottom of the classical range.

We can also graph an economy using two separate AS curves: one for the short run and one for the long run *(see Diagram 26)*. To show an economy operating at full employment, the aggregate demand curve (representing the total demand for final goods and services) and short-run aggregate supply curve (representing the total supply of goods and services produced) intersect at the same point of the **long-run aggregate supply curve**. I suggest using this approach when drawing your AD/AS graphs on the examination because it allows for greater flexibility.

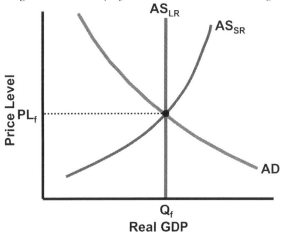

In the long run, the economy produces at its ***natural rate of unemployment***

To illustrate an **economy in recession,** the aggregate demand curve must intersect the short-run AS curve before the vertical long-run AS curve *(see Diagram 27).* Another way of illustrating a recession is making AD intersect the 3-segmented AS curve to the left of the classical range *(see Diagram 27a).*

Recession in the Short Run and Long Run: In the short run, fiscal and monetary policies can be used to shift AD right, toward full-employment. In the long run–assuming no policy is enacted–wages will eventually fall and shift the short-run AS curve to the right. This will reestablish the full-employment level of output.

Diagram 27: Recession with Short-Run and Long-Run Aggregate Supply Curves

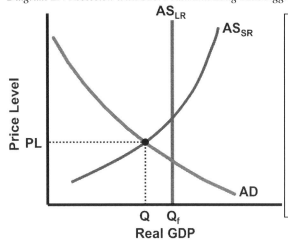

Recession
In the ***short run***, an expansionary fiscal or monetary policy can shift AD right toward full-employment

In the ***long run***, nominal wages will fall and shift AS_SR right toward full-employment

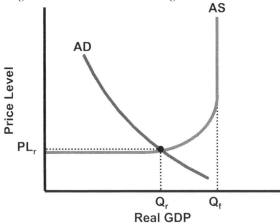

Diagram 27a: Recession with a 3-Segmented AS Curve

During a ***recession***, an increase in AD will increase real GDP and price level, and reduce unemployment

An **economy with inflation** can be shown by drawing an aggregate demand curve that is intersecting the short-run AS curve to the right of the vertical long-run AS curve *(see Diagram 28)*. This can also be shown by drawing an AD curve up in the vertical range of the 3-segmented AS curve *(see Diagram 28a)*.

Inflation in the Short Run and Long Run: In the short run, a contractionary fiscal or monetary policy can help shift AD left to reduce the price level. However, in the long run, workers will demand higher wages causing the short-run AS curve (or the Keynesian and intermediate ranges of the three-segmented AS curve) to shift left, back to full-employment output.

Diagram 28: Inflation with Short-Run and Long-Run Aggregate Supply Curves

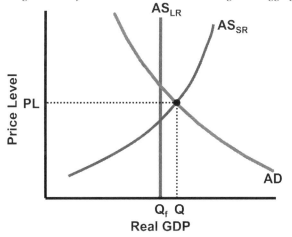

Inflation

In the ***short run***, a contractionary fiscal or monetary policy can shift AD left toward full-employment

In the ***long run***, nominal wages will rise and shift AS$_{SR}$ left toward full-employment

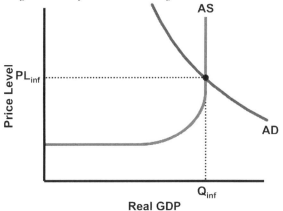

During times of high **inflation**, a decrease in AD will decrease real GDP and price level, and cause an increase in unemployment

The four **determinants or shift factors of aggregate demand** are changes in consumption, investment, government spending, and net export spending–essentially, the same components that make up GDP through the expenditures approach.

Suppose a decrease in disposable income or stock market wealth causes consumption to fall, forcing the AD curve to shift left in the short run (*see Diagram 29*). The equilibrium price level falls, output decreases, and unemployment rises. Now the economy is in recession. When drawing shifts, be sure to include arrows to indicate direction.

Diagram 29: Decrease in AD: Price Level Decreases, Output Decreases

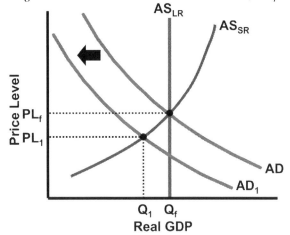

A decrease in consumption, investment, government spending, or net exports will *shift AD to the left*

Now let's look at a rightward shift in AD (*see Diagram 30*). Say there is an increase in investment spending that causes AD to increase in the short run. The result is a higher price level, increased output, and less unemployment.

Diagram 30: Increase in AD: Price Level Increases, Output Increases

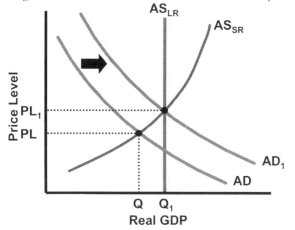

An increase in consumption, investment, government spending, or net exports will **shift AD to the right**

Here are the **shift factors of short-run aggregate supply**. These include changes in resource prices, productivity, taxes and subsidies on production, and inflation expectations.

On the graph below, you can see a classic example of a **negative supply shock** *(see Diagram 31)*; here, an increase in the price of oil–a major economic resource–causes short-run AS to shift left. The price level increases, which indicates inflation, and output decreases indicating a recession: two evils at the same time, a condition known as **stagflation**. When this happens, policy makers find themselves in a bit of a pickle because their policies focus primarily on shifting AD.

Diagram 31: Decrease in Short-Run Aggregate Supply: Price Level Increases, Output Decreases

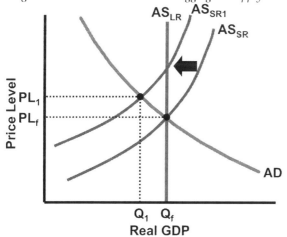

An increase in resource costs, inflation expectations, or taxes on production will **shift AS$_{SR}$ to the left**

Fiscal policy is economic policy conducted by the government to help the economy expand or contract by shifting aggregate demand in the short run. There are two **tools of fiscal policy**: changes in government spending and changes in taxes.

Expansionary Fiscal Policy: If the economy is in recession, an appropriate fiscal policy is an expansionary one. It could increase government spending, which can be a powerful **injection**

into the economy when consumption and investment are low. This directly increases AD, causing prices and output to rise, and causing unemployment to fall.

The government can also lower personal income taxes, which increases **disposable income** (income after taxes) and consumption. This also causes AD to shift right, however, it is not quite as powerful as an increase in spending–from the Keynesian perspective–because households save some of their income, and savings is a **leakage**.

Contractionary Fiscal Policy: If the economy is experiencing high prices, a contractionary policy will be more appropriate. This includes a decrease in government spending or an increase in personal income taxes. Aggregate demand will shift left, leading to a lower price level and output; however, this could cause an increase in unemployment.

Any change in **autonomous spending**, such as government spending, has a magnified effect on output. This is known as the **multiplier effect**.

The multiplier can be found using the **marginal propensities to consume and save.** Income is either spent or saved so the MPS plus the MPC is always equal to 1 *(see Diagram 32)*. If you know one, you can easily find the other. The MPC is also equal to the change in consumption divided by the change in disposable income. The MPS is the change in savings divided by the change in disposable income.

Diagram 32: MPC & MPS Formula

$$MPC + MPS = 1$$

$$MPC = \frac{\text{Change in Spending}}{\text{Change in Income}}$$

If the MPC is .75 then the MPS is .25. This means that you will spend 75% of any change in income and save 25% of any change in income. To find the **spending multiplier** *(see Diagram 33)*, simply calculate 1 divided by MPS, or 1/.25, which equals a multiplier of 4.

Diagram 33: Spending Multiplier Formula

$$\text{Multiplier} = \frac{1}{MPS} \text{ or } \frac{1}{1-MPC}$$

$$MPS = \frac{\text{Change in Savings}}{\text{Change in Income}}$$

You can use a **tax multiplier** *(see Diagram 34)*, which is equal to negative MPC divided by the MPS when there is a change in taxes as part of a fiscal policy. This formula accounts for the savings leakage mentioned earlier.

Diagram 34: Tax Multiplier Formula

$$\text{Tax Multiplier} = \frac{MPC}{MPS}$$

The *spending multiplier* is stronger than the *tax multiplier*

To determine the overall **change in output from a change in spending**, you simply multiply the initial change in spending by the multiplier *(see Diagram 35)*. The reason why there is a magnified effect on output is because one person's spending becomes another person's income, and more income leads to more spending. The spending multiplier is most powerful in the Keynesian range of the AS curve. When you are working with a change in taxes, multiply the change in taxes by the tax multiplier to get a change in output (remember, a tax increase reduces output and a tax cut will increase output).

Diagram 35: Change in Output from a Change in Spending Formula

Diagram 35: Change in Output from a Change in Spending Formula

Change in Real GDP = Change in Spending X Multiplier

Let's not forget about what was mentioned earlier about the power of government spending relative to changes in taxes. If the government increases spending by $200 million and increases taxes by $200 million, output will increase by $200 million *(see Diagram 36)*. This is because the **balanced budget multiplier** is equal to 1.

Diagram 36: Balanced Budget Multiplier

Balanced Budget Multiplier = 1

Balanced Budget ($\uparrow$G = $\uparrow$Tax)
Δ Gov't Spending = Δ Real GDP

The last model that I want to mention is the **aggregate expenditures model**, which is similar to the AD/AS model. When aggregate expenditures like C + I + G meet the 45 degree line before the full-employment level of output, a **recessionary gap** exists *(see Diagram 37)*. The AE curve must shift up to reach full employment. Do not stress yourself out about drawing the AE model on your own; instead be familiar with the terminology.

Diagram 37: Recessionary Gap in the Aggregate Expenditures Model

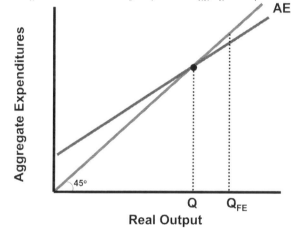

During a **recessionary gap**, the potential GDP exceeds actual GDP

If the aggregate expenditures curve meets the 45-degree line beyond full employment, an **inflationary gap** exists *(see Diagram 38)*. A decrease in spending in the short run can have a multiplied decrease in real GDP, which could restore full employment.

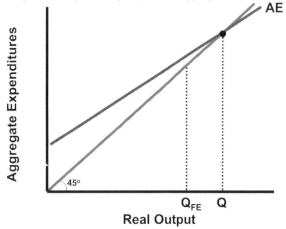

Diagram 38: Inflationary Gap in the Aggregate Expenditures Model

During an ***inflationary gap***, the actual GDP exceeds potential GDP

As far as graphing goes on the free response section of the Macroeconomics exam, you should show up on test day feeling very comfortable **drawing AD/AS graphs on your own**. Practice before test day by graphing the economy in the following ways: full employment, recession, inflation, expansionary fiscal and monetary policies, and contractionary fiscal and monetary policies.

Those are the major concepts from unit three of Macroeconomics. Go test your skills by completing the worksheet and practice questions on AD/AS & Fiscal Policy.

NB3. AD/AS & Fiscal Policy (Macro) – Worksheet

Identify the following terms:

a. Aggregate Demand _Demand at whole economy_

b. Aggregate Supply _Supply of whole economy_

c. Stagflation _high inflation, high employment_

d. Fiscal Policy _tool US govt uses to stabilize econ_

e. Marginal Propensity to Consume _consumption per ↑ income_

f. Marginal Propensity to Save _saving per ↑ income_

g. Multiplier Effect _idea that save spending gets magnified_

h. Spending Multiplier _1/mps   1-mpc_

i. Tax Multiplier _Spending -1   or   mps/mps_

j. Balanced Budget Multiplier _____

Summary questions:

1. How do you illustrate an economy in long-run equilibrium (full-employment equilibrium) using the AD/AS model? _____
   AD & AS spots (RA)

2. How do you illustrate an economy in recession using the AD/AS model? _____
   demand shifts left

3. If the economy is in recession, what will happen to short-run aggregate supply in the long run? Why? _more high prices will drop and supply will shift right_

41

4. How do you illustrate an economy experiencing inflation using the AD/AS model? ____
   _____ demand _____ _____ right _____
   _____
   _____
   _____

5. What are the shift factors of aggregate demand? _____
   _____ consumption investment govt spending _____ exp _____
   _____

6. What are the shift factors of short-run aggregate supply? _____
   _____ productivity tax/subsidies resource prices _____
   _____

7. What is an example of an expansionary fiscal policy? What are the effects of an
   expansionary fiscal policy when the economy is in recession? _____
   _____ tax/ govt spend, tax _____
   _____
   _____

8. What is a contractionary fiscal policy? What are the effects of a contractionary fiscal
   policy when the economy is experiencing high inflation? _____
   _____ Both tax ↑ gov spend _____
   _____
   _____

9. What is the difference between the spending multiplier and tax multiplier? Which is
   more powerful? _____ spend tax _____ applied to two _____
   _____
   _____

10. How do you determine the change in real GDP from a change in government spending?
    _____ ↑ Govt spend ↑ GDP _____
    _____
    _____

NB3. AD/AS & Fiscal Policy (Macro) – Practice

1. In the Keynesian model, a decrease in which of the following will lead to an increase in unemployment?
    a. Gross investment
    b. Interest rates
    c. Taxes
    d. Imports
    e. International value of the dollar

2. An increase in which of the following will cause the shift in aggregate demand shown below?

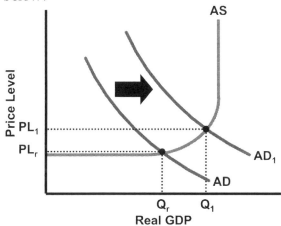

    a. Productivity
    b. International value of the dollar
    c. Personal income taxes
    d. Wages
    e. Exports

3. Which of the following will force the economy into a period of stagflation?
    a. Increase in aggregate demand
    b. Decrease in aggregate demand
    c. Increase in aggregate supply
    d. Decrease in aggregate supply
    e. None of the above

4. A decrease in which of the following will most likely cause the shift shown below?

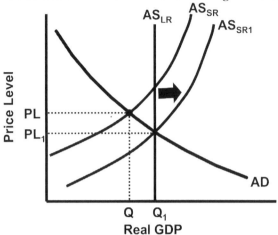

    a. Government spending
    b. Price of land
    c. Productivity
    d. Imports
    e. Exports

5. When the economy dips into a recession, the federal budget moves towards a _____ because of the presence of automatic stabilizers such as _____.
    a. deficit; unemployment compensation
    b. deficit; discretionary spending
    c. surplus; transfer payments
    d. surplus; discretionary spending
    e. surplus; welfare payments

6. Assume that disposable income increases by $2,000. If the marginal propensity to save is 0.25, consumer expenditures will increase by
    a. $500
    b. $1,000
    c. $1,500
    d. $2,000
    e. $2,500

7. As a result of a $25 billion increase in government spending, the real gross domestic product increases by $125 billion. What is the marginal propensity to consume?
    a. 0.2
    b. 0.25
    c. 0.75
    d. 0.8
    e. 1

8. Suppose that the aggregate demand curve intersects the short-run aggregate supply curve at an output level that exceeds full employment. Which of the following will occur in the long run?

I. Aggregate demand will shift right to return the economy to long-run equilibrium.

II. The long-run aggregate supply curve will shift right to return the economy to long-run equilibrium.

III. Wages will fall and shift the short-run aggregate supply curve to the right to return the economy to long-run equilibrium.

IV. Wages will rise and shift the short-run aggregate supply to the left to return the economy to long-run equilibrium.

    a. I only
    b. II only
    c. III only
    d. IV only
    e. I & IV

9. Suppose that the marginal propensity to consume is 0.8 and the government is operating with a balanced budget. If the government increases spending and taxes by $350 million, what will happen to real gross domestic output?

    a. Increase by $350 million
    b. Increase by $1,400 million
    c. Increase by $1,750 million
    d. Decrease by $350 million
    e. Decrease by $1,400 million

10. Suppose that the marginal propensity to save is 0.2 for a closed economy with lump-sum taxes. If the government increases taxes by $200 million, real GDP could decrease by a maximum of

    a. $40 million
    b. $160 million
    c. $200 million
    d. $800 million
    e. $1 billion

Free Response Practice

A: Using an **aggregate demand and aggregate supply graph**, show an economy that is operating at full employment.

B: Suppose that lower interest rates increase investment spending. Illustrate the result of this increase in investment spending on your graph. What happens to real GDP, price level, and unemployment?

C: Explain what will occur in the long run as the economy returns to its long-run equilibrium.

_____

_____

_____

_____

_____

_____

_____

_____

_____

_____

_____

_____

_____

_____

_____

_____

_____

_____

_____

_____

_____

_____

NB3. AD/AS & Fiscal Policy (Macro) – Practice Answers

1. A, Gross Investment. If a decrease in gross investment occurs, aggregate demand will decrease. The results are lower prices, decreased real GDP, and increased unemployment. If the international value of the dollar falls, exports will increase causing AD to increase. All of the other choices will increase AD as well.

2. E, Exports. An increase in exports will cause aggregate demand to shift to the right. An increase in income taxes and the international value of the dollar will shift AD left. Increases in productivity and wages will affect aggregate supply.

3. D, Decrease in aggregate supply. When aggregate supply shifts to the left, the price level increases, output falls, and unemployment rises. High prices (inflation) + high unemployment = stagflation. When you add inflation and unemployment together, you get the **Misery Index**.

4. B, Price of land. Land is an economic resource. If the price of land falls then the cost of production falls. Aggregate supply will shift right. The price level will decrease and output will increase.

5. A, deficit; unemployment compensation. When a recession occurs, unemployment rises. The government does not have to pass a new law to pay people that file for unemployment compensation. This is known as an **automatic stabilizer**. Tax revenues also decline, as people aren't earning as much income for the government to tax. When the government passes a recovery act that increases spending, it is known as **discretionary spending**.

6. C, $1,500. If the marginal propensity to save (MPS) is 0.25 then the marginal propensity to consume (MPC) is 0.75. This means that consumers will spend 75% of their disposable income ($2,000 x 0.75 = $1,500) and save 25% ($2,000 x 0.25 = $500). MPC plus MPS equals 1 because disposable income can either be spent or saved.

7. D, 0.8. The change in government spending multiplied by the multiplier is equal to the change in real GDP ($25 billion x Multiplier = $125 billion). The multiplier is equal to 5. Multiplier is equal to 1 divided by the MPS (5 = 1/MPS). If MPS is 0.2, then the MPC must be 0.8.

8. D, IV only. The economy that is described here is one that is experiencing inflation. In the long run, workers will request higher wages because of the higher prices (inflation). Higher wages, a resource cost, will shift the short-run aggregate supply curve left toward long-run equilibrium.

9. A, Increase by $350 million. If there is an equal increase in government spending and taxes, the government is operating with a balanced budget. The multiplier in this case is equal to 1. Remember, the spending multiplier is stronger than the tax multiplier. This is why real output will still increase. It will increase by an amount that is equal to the change in government spending.

10. D, $800 million. The tax multiplier is MPC/MPS (0.8/0.2 = 4). A tax increase of $200 million when multiplied by 4 will yield a decrease in output of $800 million.

Free Response Solution

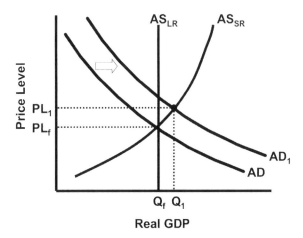

A/B: The economy above was initially in long-run equilibrium (or full employment level of output). The increase in investment spending shifts aggregate demand to the right. The price level and real GDP will increase. Unemployment will decrease.

C: In the long run, laborers will demand higher wages from their employers due to the inflation. However, higher wages will now shift the short-run aggregate supply curve to the left, intersecting the aggregate demand curve and long-run aggregate supply curve. The new price level will be higher and output will decrease.

NB4. Banking & Monetary Policy (Macro) – Review

In this chapter, we will review the key ideas of money, banking, and monetary policy.

To begin, let's look at the contents that comprise the **money supply** using the "M"s (*see Diagram 39*).

M1 is the most liquid definition of money. It contains fiat currency, checkable deposits, and traveler's checks. This best fits the **medium of exchange** function of money.

M2 serves as the **store of value** function of money, and contains everything in M1 plus non-checkable savings deposits and small time deposits that are less than $100,000.

M3 serves as the **unit of account** or standard of value function. It contains everything in M2 plus large time deposits and institutional money market funds.

Diagram 39: The "M"s

The "M"s	Contents
M1	Currency + Checkable Deposits + Traveler's Checks
M2	M1 + Savings + Small Time Deposits
M3	M2 + Large Time Deposits + Institutional Money Market Funds

Checkable deposits are also known as ***demand deposits***, which are recorded as ***liabilities*** on a bank's ***balance sheet***

Depositing ***cash*** into a checking account does not change the size of ***M1***

Time Value of Money Formulas

$$\text{Future Value} = \text{Present Value} (1 + \text{Interest Rate})^{\text{Years}}$$

$$\text{Present Value} = \text{Future Value} / (1 + \text{Interest Rate})^{\text{Years}}$$

Banks are vital to the expansion of the money supply and need to carefully track all transactions. This can be done using a **balance sheet**, which records the assets, liabilities, and net worth of the institution (*see Diagram 40*).

Checkable deposits, also known as **demand deposits** are subject to a reserve requirement that is set by the **Federal Reserve**, the central bank of the United States. The rest, or **excess reserves**, can be lent out to households and businesses. Money is created when banks make loans.

In the bank balance sheet shown below, people deposited $20,000 into their checking accounts. These demand deposits are **liabilities** because the bank must pay depositors when withdrawals are made. Also, if depositors write checks or pay for goods with their debit cards, the banks are obligated to pay the payee.

Required reserves represent a percentage of demand deposits that banks must keep on reserve. In this example, the reserve ratio is 15 percent ($3,000 is 15% of $20,000). This bank can lend out a total of $17,000 from the $20,000 of demand deposits. Since this bank already issued $12,000 worth of loans, it can lend out an additional $5,000. Reserves and loans are **assets** to the bank.

Diagram 40: Bank Balance Sheet

Assets	Liabilities
Required Reserves $3,000	Demand Deposits $20,000
Excess Reserves $5,000	
Loans $12,000	

> The **reserve ratio** indicates the percentage of **demand deposits** that must be kept as **required reserves**

The expansion of the money supply from a single deposit in a checking account is magnified due to the **money multiplier** *(see Diagram 40a)*. The multiplier is equal to one divided by the reserve requirement. If the reserve requirement is 10% then the money multiplier is 10.

Diagram 40a: Money Multiplier Formula

$$\text{Money Multiplier} = \frac{1}{\text{Reserve Ratio}}$$

> Know the four money supply formulas in NB4

To calculate the **change in the money supply from a deposit** to a checking account, simply multiply the excess reserves from the deposit by the money multiplier.

Suppose your best friend deposits $100 into his or her checking account instead of buying you an awesome birthday gift. If the reserve ratio is 10%, the bank will keep $10 while excess reserves will increase by $90. Therefore the change in the money supply from the $100 initial deposit is $90 multiplied by 10, or $900.

To calculate **the change in checkable deposits of the banking system**, multiply the initial deposit by the multiplier.

The **money market** shows how changes in the demand and supply of money affect **nominal interest rates**. The change in interest rates impacts the level of investment spending and aggregate demand, bond prices, and the international value of a currency.

If the **demand for money shifts right** because consumers demand more cash for transactions, the nominal interest rate rises, and bond prices will fall *(see Diagram 41)*. The money supply is vertical because the Federal Reserve controls the supply of money in the economy.

Diagram 41: Money Demand Increases in Money Market: Nominal Interest Rate Increases

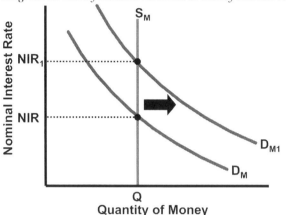

As people demand more cash to increase spending, *nominal interest rates* will increase and *bond prices* will fall

If the **demand for money shifts left**, the nominal interest rate falls, and bond prices rise *(see Diagram 42)*. Note the **inverse relationship between bond prices and interest rates**.

Diagram 42: Money Demand Decreases in Money Market: Nominal Interest Rate Decreases

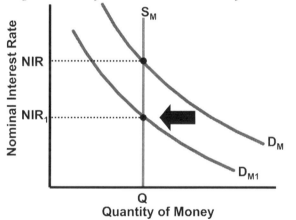

As people cut back on cash holdings and spending, *nominal interest rates* will decrease and *bond prices* will increase

Monetary policy is conducted by the Fed to promote growth and to keep prices stable. This is done with **open market operations**, which includes the buying and selling of government bonds or securities. It also includes changing the **discount rate**, which is the interest rate that the Fed charges banks for overnight loans. Another tool of the Fed is to change the **reserve requirement**.

During a recession, the Fed should pursue an **easy monetary policy**. Here, the Fed will buy bonds on the open market; however, the Fed also has the options to lower the discount rate and/or lower the reserve requirement. All three options will increase excess reserves of the banking system, but the Fed prefers to target the **federal funds rate**, which is the bank-to-bank interest rate for overnight loans, using open market operations.

When the Fed buys treasury bonds, the money supply shifts right in the money market *(see Diagram 43)*. This monetary policy action causes the nominal interest rate to fall. **To determine** the overall change in money supply from an open market operation, multiply the bond purchase by the money multiplier.

Diagram 43: Easy Monetary Policy: Money Supply Increases: Nominal Interest Rate Decreases, Quantity Increases

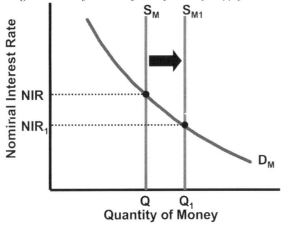

When the Fed **buys bonds**: **excess reserves** rise, money supply shifts to the right, and the nominal interest rate decreases

This is most appropriate during a **recession**

Expansionary Monetary Policy: The chain of events for an easy monetary policy is as follows: buying government bonds will increase the money supply and lower interest rates. This causes consumption and investment spending to increase. Aggregate demand will shift right, causing the price level and real GDP to increase. Therefore, the unemployment rate will decrease *(see Diagram 44)*. You will see in the final unit of Macroeconomics that the lower interest rate can also lead to the depreciation of the currency causing net exports to increase.

Diagram 44: Keynesian Transmission Mechanism (Easy Monetary Policy)

Inc. MS, Dec. I.R., Inc. C & Ig, Inc. AD, Inc. PL, Inc. RGDP, Dec. Unemployment

A lower interest rate will also reduce foreign demand for currency and **depreciate** its value

During an inflationary period, a **tight monetary policy** is most appropriate. The Fed can sell bonds, which will raise the federal funds rate, increase the discount rate, and/or increase the reserve requirement.

When the Fed sells government securities, the money supply in the money market shifts left *(see Diagram 45)*, resulting in a higher nominal interest rate.

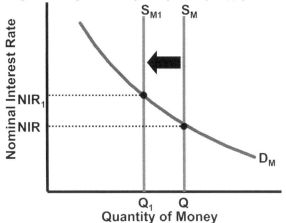

When the Fed **sells bonds**: **excess reserves** fall, money supply shifts to the left, and the nominal interest rate increases

This is most appropriate during **high inflation**

Contractionary Monetary Policy: The chain of events for a tight monetary policy is as follows: selling government bonds will decrease the money supply thus raising interest rates. Consumption and investment spending will fall, causing aggregate demand to shift left. The price level and real GDP will decrease, while the rate of unemployment will increase *(see Diagram 46)*. You will see in the final unit of Macroeconomics that higher interest rates can lead to greater international demand for a nation's currency, which causes its appreciation and a decrease in net exports.

Diagram 46: Keynesian Transmission Mechanism (Tight Monetary Policy)

Dec. MS, Inc. I.R., Dec. C & Ig, Dec. AD, Dec. PL, Dec. RGDP, Inc. Unemployment

A higher interest rate will also increase foreign demand for currency and **appreciate** its value

Those are the major concepts from unit four of Macroeconomics. Go test your skills by completing the worksheet and practice questions on Money & Monetary Policy. Following the practice questions and short graphing practice, you will find the answers and explanations. In NB5, we will review the crowding out effect, the equation of exchange, rational expectations theory, the Phillips curve, and the key determinants of economic growth.

NB4. Banking & Monetary Policy (Macro) – Worksheet

Identify the following terms:

a. Balance Sheet _____

b. Demand Deposit _____

c. Required Reserves _____

d. Excess Reserves _____

e. Money Multiplier _____

f. Money Market _____

g. Monetary Policy _____

h. Open Market Operations _____

i. Discount Rate _____

j. Federal Funds Rate _____

Summary questions:

1. What are three functions of money? _____
   _____
   _____

2. What is included in the M1 definition of money? M2? M3? _____
   _____
   _____

3. How do you determine the change in money supply from a new checkable deposit? _____
   _____
   _____

4. How do you determine the change in overall checkable deposits in the banking system from a new demand deposit? _____
   _____
   _____

5. Why is the money supply depicted as a vertical line in the money market? _____
   _____
   _____

6. How does an increase in demand for money affect nominal interest rates and bond prices? _↑ , ↓_

7. What are three tools that monetary policy makers have to alter the money supply? ____
   fed reserve ratios, bonds, discount rate

8. What are appropriate monetary policy actions when the economy is experiencing a recession? Appropriate policy actions for inflation? _buy bonds, sell bonds,_
   gov spend, int ↓

9. What will happen to the money supply, interest rates, investment spending, aggregate demand, price level, real GDP, and unemployment when the Fed buys treasury bonds?
   all up, unemploy↓

10. What will happen to the money supply, interest rates, investment spending, aggregate demand, price level, real GDP, and unemployment when the Fed sells treasury bonds?
    all down, unemploy↑

NB4. Banking & Monetary Policy (Macro) – Practice

1. Which of the following is not part of M2?
    a. Checkable deposits
    b. Savings deposits
    c. Traveler's checks
    d. Small time deposits
    e. Large time deposits

2. Fiat money _____.
    I. has no intrinsic value
    II. is legal tender
    III. is a medium of exchange
    IV. is a store of value
    a. II only
    b. II and III
    c. III and IV
    d. II, III, and IV
    e. I, II, III, and IV

3. Assume that the demand for money increases in the money market. If the supply of money remains constant, what will happen to the equilibrium interest rate and quantity?
    a. Interest rate increases; Quantity increases
    b. Interest rate increases; Quantity stays the same
    c. Interest rate decreases; Quantity decreases
    d. Interest rate decreases; Quantity stays the same
    e. Interest rate increases; Quantity decreases

4. A decrease in which of the following could have caused the shift shown below?

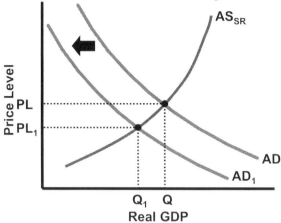

a. Discount rate
b. Imports
c. Open market purchases of government bonds
d. Reserve requirement
e. Personal income taxes

5. If the Fed _____ bonds on the open market, the nominal interest rate will _____ and bond prices will _____.
a. sells; decrease; decrease
b. buys; decrease; decrease
c. sells; decrease; increase
d. buys; decrease; increase
e. sells; increase; increase

6. Suppose that the required reserve ratio is 10% and banks lend out all of their excess reserves. Ross the Boss earned $1,500 cash yesterday and decides to deposit $1,000 of his earnings in his checking account while stashing the other $500 under his mattress. What will happen to the total amount of checkable deposits in the banking system?
a. Increase by $5,000
b. Increase by $9,000
c. Increase by $10,000
d. Increase by $13,500
e. Increase by $15,000

7. The shift shown below is a result of the Federal Reserve _____ bonds on the open market. As a result of this policy, investment spending will _____.

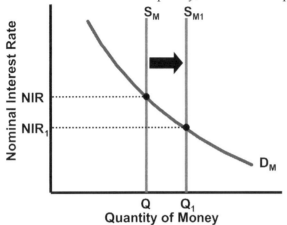

a. buying; increase
b. selling; increase
c. selling; decrease
d. buying; decrease
e. buying; stay the same

8. The interest rate at which banks make short term loans to one another is commonly known as the
a. prime rate
b. discount rate
c. federal funds rate
d. nominal rate
e. annual percentage rate

9. Assume that the reserve requirement is 10% and the Federal Reserve buys $5 billion worth of government bonds from the public. The maximum increase in the money supply is
a. $45 billion
b. $50 billion
c. $100 billion
d. $450 billion
e. $500 billion

10. Assume that the economy is operating at a level that is less than full employment. If the Federal Reserve buys government bonds, how will output and real wages change in the short run?
a. Output increases; Real wages increase
b. Output increases; Real wages decrease
c. Output decreases; Real wages decrease
d. Output decreases; Real wages increase
e. Output increases; Real wages stay the same

Free Response Practice

A: If the economy is currently experiencing high demand-pull inflation, identify the best open market operation that the Federal Reserve should undertake.

B: Using a **money market graph**, illustrate the monetary policy action you identified above. What will happen to the nominal interest rate, the quantity of money, and the unemployment rate? Explain.

sell bond

p

NB4. Banking & Monetary Policy (Macro) – Practice Answers

1. E, Large time deposits. M2 includes everything in M1 plus savings deposits and small time deposits. Large time deposits (over $100,000) will fall in M3.

2. E, I, II, III, and IV. Fiat money would have no intrinsic value. It has value because the government declares that it is legal tender. It can be used as a medium of exchange, a store of value, or a standard of value (unit of account).

3. B, Interest rate increases, Quantity stays the same. In the money market, the money supply is set by the Fed. If demand for money increases, interest rates rise and the quantity of money stays constant.

4. C, Open market purchases of government bonds. If the Federal Reserve decreases its open market purchases of government securities then aggregate demand will shift to the left. This is because the Fed's policy will increase interest rates causing consumption and investment to fall.

5. D, buys; decrease; increase. When the Fed buys government bonds, the money supply shifts to the right in the money market and causes the nominal interest rate to decrease. Interest rates and bond prices are inversely related, so bond prices will increase when the nominal interest rate decreases.

6. C, Increase by $10,000. The money multiplier is 1 divided by the reserve ratio ($1/0.1 = 10$). The total increase in demand deposits is the initial deposit of $1,000 multiplied by the money multiplier of 10.

7. A, buying; increase. When the Fed buys bonds, the money supply shifts right. This causes the nominal interest rate to fall and private spending to increase. As a result of this easy monetary policy, aggregate demand will increase.

8. C, federal funds rate. The federal funds rate is the interest rate for overnight loans that banks charge one another. This interest rate affects other key interest rates within the economy (mortgage rates, auto loans, school loans, etc). The Fed indirectly sets this rate by buying and selling government bonds on the open market. When the Fed buys bonds, the federal funds rate falls and vice versa.

9. B, $50 billion. The money multiplier is 1 divided by the required reserve ratio ($1/0.1 = 10$). Multiply the money multiplier by the $5 billion purchase of securities ($5 billion x 10 = $50 billion) to get the maximum increase in the supply of money.

10. B, Output increases; Real wages decrease. If the Fed buys bonds, the money supply increases causing interest rates to fall. Aggregate demand shifts right, and both the output and price level will rise. Real wages will fall due to the inflation brought on by the expansionary monetary policy. In other words, nominal wages will have less purchasing power.

Free Response Solution

A: The appropriate open market operation is to sell bonds.

B: This will reduce excess reserves in the banking system and shift the supply of money to the left in the money market. The nominal interest rate will increase and the quantity of money will decrease. Investment spending will fall which decreases aggregate demand. The price level and real GDP will decrease while unemployment will increase.

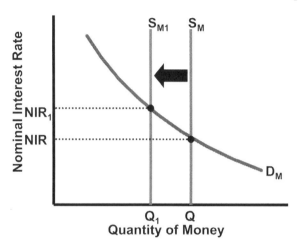

NB5. Policies & Growth (Macro) – Review

In this section, we will continue to discuss the main ideas behind economic policies and theories, as well as the factors that can cause economic growth.

Let's begin by exploring the **loanable funds market**, a model that demonstrates the relationship between the real interest rate and the quantity of loanable funds. This model is used to explain how real interest rates are altered by fiscal policy actions, changes in household savings, and international investment decisions.

We can use this model to illustrate the **crowding out effect**, a complication that is associated with **expansionary fiscal policy**.

How the crowding out effect works: When the government increases spending or reduces taxes, it must borrow funds–by issuing bonds–to finance its policy. The government runs a budget deficit. When the government borrows money from the private sector, there is an increase in the overall demand for loanable funds *(see Diagram 47)*. As a result of the government's borrowing, the real interest rate and equilibrium quantity of loanable funds will increase. The higher real interest rate will lead to a reduction (crowding out) of private sector spending (consumption and investment). This is an unintended consequence of expansionary fiscal policy.

Diagram 47: Loanable Funds: Increase in Demand from Expansionary Fiscal Policy: Real Interest Rate Increases, Quantity Increases

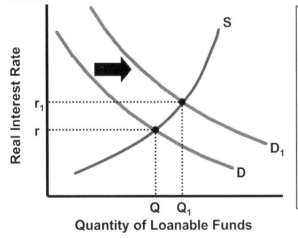

> ***Budget deficits*** from expansionary fiscal policies cause increased ***real interest rates*** and hurt long-run economic growth
>
> A higher real interest rate also causes currency to ***appreciate***, which reduces ***net exports***

An alternative way to view expansionary fiscal policy is that the **government reduces the private supply of loanable funds** *(see Diagram 48)*. This too, will lead to a higher real interest rate as well as less consumption and investment spending. Less investment spending due to the high real interest rate can be detrimental to long-term economic growth.

Diagram 48: Loanable Funds: Decrease in Private Supply from Expansionary Fiscal Policy: Real Interest Rate Increases, Quantity Decreases

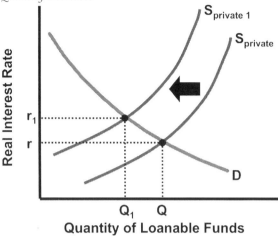

The government borrows funds by ***issuing bonds***

This reduces the private sector's supply of ***loanable funds*** and raises the real interest rate

You will see in the next chapter that the higher real interest rate can appreciate the international value of the dollar.

On the other hand, a **contractionary fiscal policy** means the government does not need to borrow as much money. This will reduce the demand for loanable funds (or increase the supply of private loanable funds), which lowers the real interest rate. This in turn can increase private spending which was not the original intention of the fiscal policy.

Diagram 49: Loanable Funds: Decrease in Demand from Contractionary Fiscal Policy: Real Interest Rate Decreases, Quantity Decreases

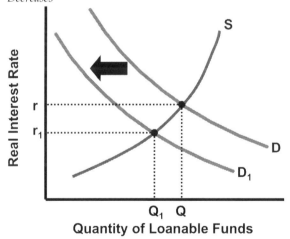

A ***budget surplus*** causes a lower ***real interest rate*** and increases long-run economic growth

A lower real interest rate also causes currency to ***depreciate***, which increases ***net exports***

When households decide to save more, there are more loanable funds available for the private sector to borrow *(see Diagram 50)*. This increased supply will reduce the real interest rate, which will increase private consumption and investment expenditures. If households reduce their savings, then the supply will shift to the left.

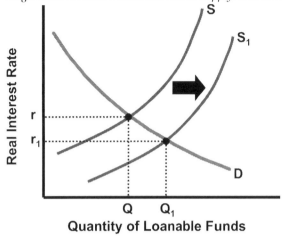

An increase in **savings**, lowers the real interest rate and encourages **investment spending**

Capital stock grows and the long-run **economic growth** rate increases

The **Phillips curve** illustrates the short-run tradeoffs between inflation (on the vertical axis) and unemployment (on the horizontal axis). The **short-run Phillips curve** is downward sloping *(see Diagram 51)*.

In the long run, there is no tradeoff between inflation and unemployment so the **long-run Phillips curve** is vertical at the natural rate of unemployment. In this example the natural rate of unemployment is 5%.

Diagram 51: Short-Run and Long-Run Phillips Curve

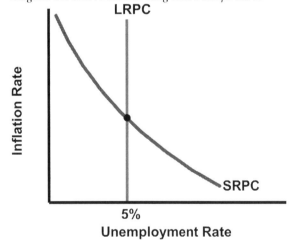

The **long-run Phillips curve** is vertical at the **natural rate of unemployment**

No tradeoff between inflation and unemployment in long run

Whenever there is an increase in aggregate demand, prices will rise and unemployment will fall. This can be shown with point-to-point movement on the short-run Phillips curve. If aggregate demand increases, movement along the short-run Phillips curve will go from point A to point B *(see Diagram 52)*. If aggregate demand shifts left, movement along the Phillips curve would go from point B to point A. Remember, a shift in aggregate demand does not shift the SRPC.

Diagram 52: Short-Run Phillips Curve: Movement from an Increase in AD

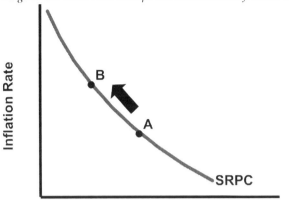

A shift of the ***aggregate demand*** curve causes ***point-to-point*** movement along the ***short-run Phillips curve*** in the opposite direction

The short-run Phillips curve will shift when the short-run aggregate supply curve shifts. If higher resource costs shift SRAS to the left, then prices and unemployment both increase *(see Diagram 53)*. This can be shown with a rightward shift of the short-run Phillips curve. Note that these curves shift in opposite directions.

Diagram 53: SRPC Shifts Right from a Decrease in SRAS

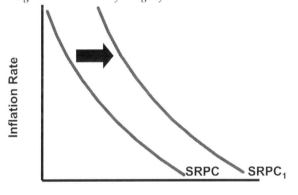

A shift of the ***short-run aggregate supply*** curve causes a ***shift*** of the ***short-run Phillips curve*** in the opposite direction

When it comes to the big disputes in Macroeconomics, we cannot ignore the battle between the Classical and Keynesian schools of thought. **Classical economists** strongly believe in the power and efficiency of markets and laissez-faire economic policy (very little government intervention). They prefer to focus on the long run and the economy's ability to self-correct. According to classical theorists: prices and wages are flexible, and more savings leads to more investment.

Keynesians argue in favor of active fiscal and monetary policies that focus on the short run. Keynesians argue that more savings during a recession leads to less consumption (the paradox of thrift), and argue that prices and wages are "sticky" in a downward direction.

Monetarists are proponents of the **monetary equation of exchange**: MV = PQ *(see Diagram 54)*. "M" represents the stock of money or money supply, "V" is the velocity of money, "P" is the price level, and "Q" is output. "P" times "Q" is therefore the nominal GDP. **Velocity** is the average amount of times a dollar is spent.

Diagram 54: Monetary Equation of Exchange

$$MV = PQ$$

Nominal GDP is equal to PQ

Monetarists charge that policies of the Federal Reserve often result in inflation, and do not change the real GDP. According to the policy concept of **monetary rule**, the Fed should only increase the money supply at a fixed rate that is equal to the expected rate of GDP growth.

The **neutrality of money** and the **quantity theory of money** are other terms associated with the ineffectiveness of monetary policy as the Fed attempts to alter real output.

Another argument assumes that people make their economic decisions with **rational expectations**. This theory also suggests that monetary policy will not change real output. Fiscal and monetary policies that seek to shift aggregate demand to increase real output are viewed as ineffective from this perspective.

Another type of economic policy focuses on the aggregate supply curve. **Supply-side economics** argues that tax incentives to businesses can lead to an increase in aggregate supply. In theory, this can be an effective way of battling stagflation.

Economic growth is an increase in real GDP or **real GDP per capita** (Real GDP/Population) over time. Growth can be caused by the following factors that increase a nation's **capital stock** (all of the capital goods within an economy): investments in research and development projects; increases in productivity; advancements in technology; more economic resources; better economic resources; training and education that leads to improvements in **human capital**.

There are two ways to illustrate economic growth. One way to show growth is with a rightward shift of the long-run aggregate supply curve (*see Diagram 55*).

Diagram 55: Rightward Shift of Long-Run Aggregate Supply

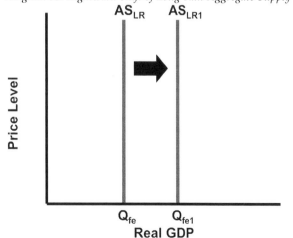

Causes of long-run growth:
1. Increased quantity of resources
2. Improved quality of resources
3. Improved education and training programs
4. Increased productivity and new technology (*research and development*)

The other way is by an outward shift of the production possibilities curve *(see Diagram 56)*. Make sure that you remember these two curves have this link.

Diagram 56: Outward Shift of the Production Possibilities Curve

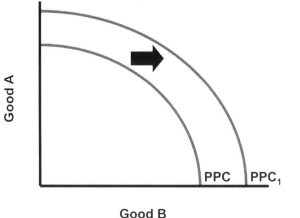

Those are the major concepts from unit five of Macroeconomics on Policies & Growth. In NB6, we will complete our review of Macroeconomics by going over the International Sector. We will summarize international trade, balance of payments, and the foreign exchange market. First, check out the nifty interest rate chart below and complete all of the NB5 practice questions.

Bonus Diagram: Nifty Interest Rate Chart

Interest Rates Increase	Interest Rates Decrease
Consumption & investment spending decrease	Consumption & investment spending increase
Aggregate demand decreases	Aggregate demand increases
Bond prices decrease	Bond prices increase
Growth of capital stock slows	Growth of capital stock increases
Long-run economic growth slows	Long-run economic growth increases
International demand for bonds & currency increase	International demand for bonds & currency decrease
Currency appreciates	Currency depreciates
Net exports decrease	Net exports increase

NB5. Policies & Growth (Macro) – Worksheet

Identify the following terms:

a. Loanable Funds Market _____

b. Crowding Out Effect _____

c. Short-Run Phillips Curve _____

d. Long-Run Phillips Curve _____

e. Monetary Equation of Exchange _____

f. Monetary Rule _____

g. Rational Expectations _____

h. Supply-Side Economics _____

i. Economic Growth _____

j. Capital Stock _____

Summary questions:

1. How do you illustrate the crowding out effect of expansionary fiscal policy using the loanable funds market? _____
    _____
    _____
    _____
    _____
    _____

2. What is the relationship between the real interest rate and investment? Real interest rate and economic growth? _____
    _____
    _____

3. What is the relationship between inflation and unemployment in the short run? _____
    _____
    _____

4. How does an increase in aggregate demand relate to the short short-run Phillips curve?
    _____
    _____
    _____

5. How does a decrease in short-run aggregate supply relate to the short-run Phillips curve? _____
_____
_____

6. What is the relationship between inflation and unemployment in the long run? _____
_____
_____

7. How do the ideas of classical economists and Keynesian economists differ? _____
_____
_____

8. List four economic terms or theories that are critical of the Federal Reserve on the basis that monetary policy cannot change real GDP: _____
_____
_____

9. What are several factors that lead to economic growth? _____
_____
_____

10. What are two ways to illustrate economic growth using the production possibilities curve and long-run aggregate supply? _____
_____
_____
_____
_____
_____

NB5. Policies & Growth (Macro) – Practice

1. The shift shown below could have been caused by a(n) _____ in government spending and a(n) _____ in the required reserve ratio.

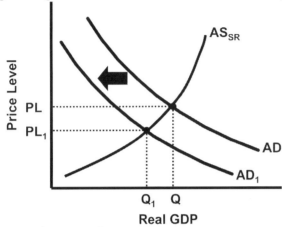

    a. increase; increase
    b. decrease; decrease
    c. increase; decrease
    d. decrease; increase
    e. decrease; indeterminate change

2. Which of the following most likely caused the shift in the loanable funds market that is shown below?

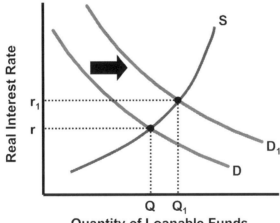

    a. Increase in government spending
    b. Decrease in government spending
    c. Increase in taxes
    d. Increase in household savings
    e. None of the above

3. Which of the following represents an appropriate fiscal policy and monetary policy to battle inflation and limit upward pressures on prices?
    a. Decreasing both spending and the discount rate
    b. Increasing taxes and selling government bonds
    c. Decreasing taxes and buying government bonds
    d. Increasing both spending and the federal funds rate
    e. Buying government bonds and decreasing the discount rate

4. Which of the following will most likely occur if the government increases spending and its budget moves from a surplus to a deficit?
    a. Aggregate supply will increase
    b. Aggregate demand will decrease
    c. Disposable income will decrease
    d. Net exports will increase
    e. Interest rates will increase

5. What is the natural rate of unemployment based on the Phillips curves shown below?

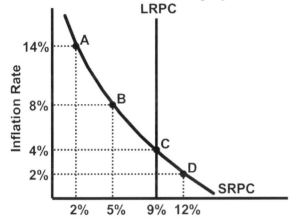

    a. 2%
    b. 5%
    c. 9%
    d. 12%
    e. Cannot be determined

6. If there is an increase in capital stock, then the
    a. production possibilities curve will shift right
    b. aggregate demand will shift left
    c. short-run Phillip's curve will shift right
    d. long-run aggregate supply will shift left
    e. economic resources of the economy will decrease

7. Which of the following most likely caused the shift in long-run aggregate supply shown below?

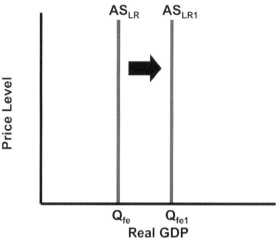

   a. Depreciation of the dollar
   b. An increase in exports
   c. An increase in technology
   d. An increase in consumption expenditures
   e. An increase in government spending

8. Suppose that prices had fallen, but velocity and money stock did not change. Which of the following must be true?
   a. Real output increased
   b. Real output decreased
   c. Real output remained the same
   d. Aggregate supply shifted left
   e. Aggregate demand shifted left

9. Monetarists argue that Federal Reserve policy makers should increase the money supply by a fixed rate that matches expected changes in
   a. the price level
   b. the international value of the dollar
   c. natural rate of unemployment
   d. real GDP
   e. velocity

10. According to the theory of rational expectations, when the Federal Reserve increases the supply of money
   a. real output stays the same
   b. unemployment increases
   c. the inflation rate falls
   d. unemployment decreases
   e. the short-run Phillips curve shifts left

Free Response Practice

A: Draw a **short-run Phillips curve** for an economy. Suppose the economy is currently operating at a level of output where the unemployment rate is 11 percent and the inflation rate is 2 percent. Label this point "R" on your graph. Assuming that the natural rate of unemployment is 4.9 percent, draw a **long-run Phillips curve** on the same graph.

B: Suppose the government decided to implement a fiscal policy that fights the high unemployment rate. Identify the fiscal policy action and assume that the action will increase the federal budget deficit.

C: Draw a **loanable funds market** that shows how this fiscal policy action will alter the real interest rate. Identify the relationship between the change in real interest rate and the rate of economic growth.

_____

_____

_____

_____

_____

_____

_____

_____

_____

_____

_____

_____

_____

_____

_____

_____

_____

_____

_____

_____

_____

NB5. Policies & Growth (Macro) – Practice Answers

1. D, decrease; increase. A decrease in government spending will decrease aggregate demand. An increase in the reserve requirement will reduce excess reserves in the banking system. Banks will make fewer loans and aggregate demand will decrease.

2. A, Increase in government spending. An increase in government spending, with no change in taxes, means that the government increases borrowing. The government is deficit spending. This shifts the demand for loanable funds to the right, which causes a higher real interest rate. This higher interest rate will crowd out private spending by households and businesses.

3. B, Increasing taxes and selling government bonds. Increased taxes (fiscal policy) and selling government securities on the open market (monetary policy) will reduce aggregate demand and lower the price level.

4. E, Interest rates will increase. When the government engages in deficit spending, it borrows money in the loanable funds market. This increased demand (or decreased private supply) raises the real interest rate. This could lead to an increase in the international value of the dollar and therefore reduce the economy's net exports.

5. C, 9%. The natural rate of unemployment is where the long-run Phillips curve meets the horizontal axis. For the economy illustrated, this occurs at an unemployment rate of 9%.

6. A, production possibilities curve will shift right. An increase in capital stock is another way of describing economic growth, which can be shown with an outward shift of the PPC.

7. C, an increase in technology. A rightward shift of long-run aggregate supply indicates economic growth. An increase in technology is a major factor of economic growth and therefore LRAS.

8. A, Real output increased. According to the equation of exchange money stock times velocity is equal to price level times real output (MV = PQ). If MV remained constant while P decreased, then Q must have increased.

9. D, real GDP. Monetarists call for fixed increases in the monetary base to match the expected growth rate of real GDP (say 3-5% annually).

10. A, real output stays the same. Rational expectation theorists would argue that an increase in the money supply increases inflationary expectations. Households buy more now (AD shifts right), workers demand higher wages (AS shifts left). Real output and employment stay constant, but prices increase.

Free Response Solution

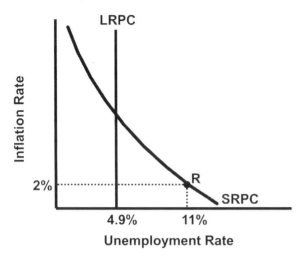

A: The economy is experiencing a recession because the current rate of unemployment is greater than the natural rate. In the short run, there is a tradeoff between inflation and unemployment, which is why the Phillips curve is downward sloping. The long-run Phillips curve is vertical at 4.9 percent unemployment because that is the natural rate of unemployment.

B: An appropriate fiscal policy action is to increase spending or cut personal income taxes, which will increase the federal budget deficit and reduce unemployment in the short run.

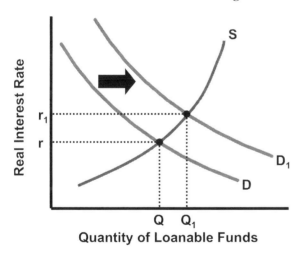

C: The demand for loanable funds will shift right (or the supply of private loanable funds will shift left) leading to higher real interest rates. The increase in real interest rates will harm private investment (crowding out effect) and slow the growth of capital stock and economic growth.

NB6. International Sector (Macro) – Review

In this chapter, we will finish our review of Macroeconomics. The main ideas behind international trade, balance of payments, and the foreign exchange market will be discussed.

You should recall from unit one, a major concept that applies to specialization and trade: **comparative advantage**. According to this law, a country with a lower relative opportunity cost of production should specialize in the production of that particular good.

In this example, Surf Kingdom has the comparative advantage in beach ball production and Sand Land has the comparative advantage in ice cream production (*see Diagram 57*). For trade to take place, both countries must benefit. Surf Kingdom must import more than 1 ice cream cone for each beach ball it exports, and Sand Land must import more than ½ a beach ball for each ice cream cone it exports. 1 beach ball for 1.5 ice cream cones would be an acceptable terms of trade.

Diagram 57: Review of Comparative Advantage

	Beach Balls	Cost of 1 Beach Ball	Ice Cream Cones	Cost of 1 Ice Cream Cone
Surf Kingdom	50	1 Ice Cream Cone	50	1 Beach Ball
Sand Land	25	2 Ice Cream Cones	50	½ Beach Ball

Terms of trade is beneficial for a country when:

Imports > Opp. Cost of Export
Exports

Make sure you can create a chart like the one above when given constant cost production possibilities curves. Also, be sure that you do not confuse comparative advantage (lower opportunity cost) with **absolute advantage** (who produces more).

Economists are often supporters of free trade because the gains outweigh the losses in the long run. **Trade barriers** such as import quotas (legal limits on imported goods) and protective tariffs (taxes on imports) are generally frowned upon. However in the short run, free trade can lead to unemployment as domestic industries face foreign competition.

The **balance of payments** system records a country's transactions with the rest of the world. A **credit** occurs when money flows into a country and a **debit** occurs when money leaves a country. The two major accounts are the current account and the financial account (capital account). In this system, a debit in the current account can be returned as a credit to the capital account, and vice versa.

The **current account** includes transactions such as exports, which credit the current account, and imports, which are recorded as debits. Net factor income and transfer payments between families abroad are also included.

The **financial (capital account)** includes investment in **real assets** such as property or manufacturing plants. The capital account also includes the sale of **financial assets** like stocks and bonds.

Suppose the Chinese buy US treasury bonds. Those purchases would credit the US financial (capital) account and contribute to a financial (capital) account surplus. When Americans buy Chinese goods, those imports debit the US current account and contribute to a current account deficit (shifting the balance of trade toward deficit as well). In theory, the accounts should eventually balance out.

The **foreign exchange market** is where exchange rates are determined by using basic supply and demand. Let's look at some of the factors that can cause a currency to strengthen or **appreciate** in value relative to another currency.

Here are the **factors that lead to the appreciation of currency**: an increase in tastes or preferences for a country's goods; higher relative interest rates; lower relative prices; a decrease in income; and **foreign direct investment** inflows, which are foreign purchases of real assets.

Suppose that we want to see the effects of higher real interest rates in the US relative to Europe using the market for US dollars. We label the horizontal axis with quantity of US Dollars, and the vertical axis is labeled Euro Price of the Dollar, or Euro/Dollar Exchange Rate (*see Diagram 58*).

To show the effects of the high US interest rates, we shift the demand for dollars to the right. Foreigners will demand **interest-bearing assets** like US treasury bonds and trade euros for US dollars. The dollar appreciates and the euro depreciates.

Diagram 58: Foreign Exchange: Increase in Demand for Dollars: Dollar Appreciates, Euro Depreciates

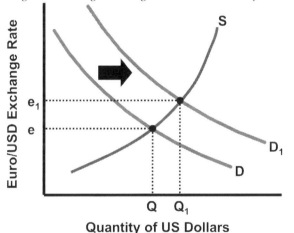

Causes of appreciation:
1. Higher interest rates
2. Increase in tastes
3. Lower inflation
4. Less income
5. More foreign direct investment

If the US goes into recession while Europe's economy remains strong, there will be a decrease in US dollars supplied to the foreign exchange market (*see Diagram 59*). This is because US citizens have less income and therefore cannot purchase as many foreign goods or foreign interest-bearing assets. The dollar appreciates and the euro depreciates.

Diagram 59: Foreign Exchange: Decrease in Supply of Dollars: Dollar Appreciates, Euro Depreciates

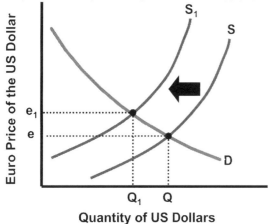

A country's ***goods look relatively more expensive*** when its currency appreciates in value

When a currency appreciates, then that country's goods will look more expensive to foreigners, and foreign goods will look relatively cheap. Exports will decrease and imports will increase, which can cause a leftward shift in aggregate demand.

Here are the **factors that lead to the depreciation of currency**: a decrease in tastes; lower relative interest rates; higher relative prices; an increase in income; and foreign direct investment outflows.

Suppose that prices in the US are higher than in Europe. Foreigners will demand fewer US goods and therefore fewer US dollars *(see Diagram 60)*. When demand for dollars shifts to the left, the dollar depreciates while the euro appreciates.

Diagram 60: Foreign Exchange: Decrease in Demand for Dollars: Dollar Depreciates, Euro Appreciates

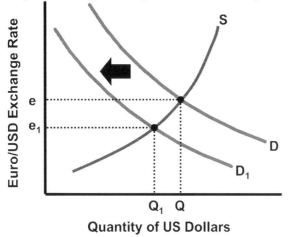

Causes of depreciation:
1. Lower interest rates
2. Decrease in tastes
3. Higher inflation
4. More income
5. Less foreign direct investment

If interest rates are higher in Europe than in the US, then people will get rid of their US dollars by exchanging them for euros *(see Diagram 61)*. The supply of dollars shifts right which causes the dollar to depreciate and the euro to appreciate.

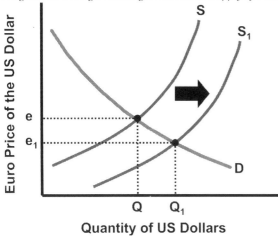

A country's **goods look relatively cheap** when its currency depreciates in value

Net exports and **aggregate demand** will increase

When a currency depreciates, then that country's goods will look relatively cheap on the international market and foreign goods will look more expensive. Exports will increase and imports will decrease, which can cause a rightward shift in aggregate demand.

It is important to apply what you learned about the AD/AS model and loanable funds market to determine what will happen to the value of a currency in the foreign exchange market.

This chapter concludes our review of Macroeconomics. Be sure that you can draw and interpret all of the graphs presented in the review chapters and practice questions.

No Bull Tip – Master these 5 Macroeconomics graphs!
1. **Aggregate Demand / Aggregate Supply**
2. **Money Market**
3. **Loanable Funds Market**
4. **Short-Run and Long-Run Phillips Curves**
5. **Foreign Exchange Market**

I recommend drawing these 5 graphs and models over and over again before exam day, including all of their variations. You should also complete all of the recent part two free-response questions that are available on the Internet for free.

Now go complete the worksheet and practice questions on the International Sector, and then take the No Bull Macroeconomics Exam. Good luck!

NB6. International Sector (Macro) – Worksheet

Identify the following terms:

    a. Comparative Advantage _____

    b. Absolute Advantage _____

    c. Trade Barriers _____

    d. Balance of Payments _____

    e. Credit _____

    f. Debit _____

    g. Current Account _____

    h. Capital (Financial) Account _____

    i. Interest-Bearing Assets _____

    j. Foreign Exchange Market _____

Summary questions:

1. How does one determine whether a country has a comparative advantage using opportunity costs? _____
_____
_____

2. What is the difference between a real asset and financial asset? _____
_____
_____

3. If the current account shows a deficit in the balance of payments system, what will the capital account show? _____
_____
_____

4. What are the factors that will lead to the appreciation of a nation's currency? _____
_____
_____

5. What is the relationship between interest rates, demand for a nation's currency, and the international value of a currency? _____
_____

6. How can a recession affect a nation's supply of currency to the foreign exchange market and the currency's international value? _____

_____

_____

7. When a nation's currency appreciates, what will happen to that nation's trade balance and aggregate demand? _____

_____

_____

8. What are the factors that will lead to the depreciation of a nation's currency? _____

_____ _____

_____

9. Suppose that the prices of goods are higher in Japan than in India. What will happen to the demand for yen and the rupee price of the yen as a result? _____

_____

_____

10. When a nation's currency depreciates, what will happen to that nation's trade balance and aggregate demand? _____

_____

_____

NB6. International Sector (Macro) – Practice

1. Economists believe in free trade because
    a. businesses don't suffer from trade in the short run
    b. long term gains surpass the long term losses
    c. domestic jobs are fully sheltered by a central governing authority
    d. people are laid off but will eventually find new jobs
    e. None of the above

2. If an American buys a DVD movie that is produced in India, the transaction would be recorded as a _____ to the US _____ account.
    a. credit; current
    b. credit; capital
    c. credit; reserve
    d. debit; capital
    e. debit; current

3. What will happen to the supply of US dollars to the foreign exchange market and the international value of the dollar based on the change in Real GDP shown below?

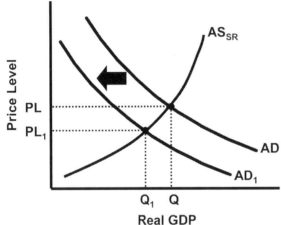

    a. Supply decreases; Dollar appreciates
    b. Supply decreases; Dollar depreciates
    c. Supply increases; Dollar appreciates
    d. Supply increases; Dollar depreciates
    e. Cannot be determined based on the information given

4. Suppose that the exchange rate between the Japanese yen and the Indian rupee are 4 yen per 1 rupee. If the exchange rate changed to 3 yen per 1 rupee then the Japanese yen has
    a. appreciated and Japan's net exports will increase
    b. appreciated and Japan's net exports will decrease
    c. depreciated and Japan's net exports will increase
    d. depreciated and India's net exports will decrease
    e. depreciated and India's net exports will increase

5. Suppose that real interest rates are lower in the United States relative to real interest rates in foreign countries. Capital would flow _____ the United States and the US dollar will _____.
    a. into; appreciate
    b. into; depreciate
    c. out of; depreciate
    d. out of; appreciate
    e. None of the above

6. What will happen to the international value of the US dollar and US net exports based on the results of the fiscal policy action shown below?

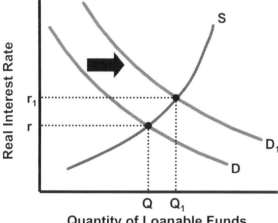

Quantity of Loanable Funds

    a. Dollar appreciates; Net exports increase
    b. Dollar depreciates; Net exports decrease
    c. Dollar depreciates; Net exports increase
    d. Dollar appreciates; Net exports decrease
    e. Cannot be determined based on the information given

7. Suppose that households increase their savings, which affects the equilibrium real interest rate in the loanable funds market. How will the change in the real interest rate affect the international value of the dollar, and how will the change in the real interest rate affect long-run economic growth?
    a. Dollar appreciates; Growth increases
    b. Dollar depreciates; Growth decreases
    c. Dollar depreciates; Growth increases
    d. Dollar appreciates; Growth decreases
    e. Cannot be determined based on the information given

8. An increase in which of the following will most likely cause the United States dollar to depreciate in the foreign exchange market?
    a. Relative price level of the United States
    b. Interest rates in the United States
    c. Demand for United States dollars
    d. Income of foreigners
    e. All of the above

9. Which of the following is true of the foreign exchange market graph shown below?

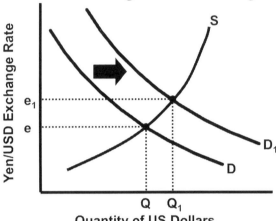

    a. The US dollar depreciated
    b. The US dollar strengthened against the yen
    c. The Japanese yen appreciated
    d. US exports will increase
    e. Yen price of the dollar decreased

10. Suppose the Federal Reserve enacts an easy monetary policy at the same time that the government enacts a contractionary fiscal policy. Based on the change in interest rates from these policies, how will the international demand for US dollars and the US long-run economic growth rate most likely change?
    a. Demand increase; Growth increase
    b. Demand decrease; Growth decrease
    c. Demand increase; Growth decrease
    d. Demand decrease; Growth increase
    e. Cannot be determined based on the information given

Free Response Practice

A: Draw a **foreign exchange market** for the euro that shows the value of the euro relative to the value of the US dollar.

B: Suppose that real income in Europe decreases relative to the real income of the United States. Explain how this change in real income will affect the supply of euros and then illustrate on your graph. What will happen to the value of the US dollar?

NB6. International Sector (Macro) – Practice Answers

1. B, Long-term gains surpass the long-term losses. According to economists, the gains to consumers and producers in the long run exceed the losses to the producers.

2. E, debit; current. A DVD movie made in India is an import and therefore a current account transaction. Because money leaves the US to pay for the DVD, it would be recorded as a debit.

3. A, Supply decreases; Dollar appreciates. A decrease in real GDP means less US income and fewer dollars for currency and foreign goods. The supply of dollars in the foreign exchange market will decrease causing the dollar to appreciate.

4. B, appreciated and Japan's net exports will decrease. It now takes fewer yen to purchase a rupee. This means that the yen appreciates in value relative to the rupee. Japan's goods will be more expensive for Indians (Japan exports less) and Indian goods will look cheaper for the Japanese (Japan imports more).

5. C, out of; depreciate. Capital will flow out of the US because lower interest rates are less attractive to international investors that want interest-bearing assets. This will depreciate the value of the dollar.

6. D, Dollar appreciates; Net exports decrease. The higher real interest rate will lead to greater demand for US interest-bearing assets and US dollars. The dollar appreciates, which makes US goods look more expensive to foreigners (exports decrease). Foreign goods will look cheaper for Americans (imports increase).

7. C, Dollar depreciates; Growth increases. The supply of loanable funds will shift right and lower the real interest rate. The lower real interest rate will encourage investment spending and increase long-run economic growth. The lower real interest rate will make the dollar look less attractive and foreigners will demand fewer US interest-bearing assets. The decreased demand for the dollar will depreciate the dollar.

8. A, Relative price level of the United States. If the prices of US goods are relatively high compared to the price levels of other countries, then there will be less foreign demand for US goods. There will be less demand for US dollars, which leads to a depreciation of the US dollar.

9. B, The US dollar strengthened against the yen. Demand for US dollars shifted right, raising the yen price of the dollar. The dollar appreciated and the yen depreciated as a result of the shift. Americans will increase imports because foreign goods will look relatively cheap. US exports will decrease.

10. D, Demand decrease; Growth increase. Easy monetary and contractionary fiscal policies result in lower interest rates. Foreigners will now demand fewer US dollars and fewer interest-bearing assets (US bonds). Businesses in the US are more likely to purchase capital goods due to the lower interest rates, which encourages economic growth.

Free Response Solution

A/B: If real income in Europe decreases, fewer euros will be supplied to the foreign exchange market. The euro will appreciate and the US dollar will depreciate. European goods will now look relatively expensive to Americans.

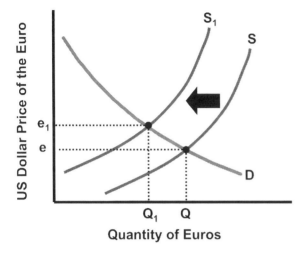

NB7. Utility & Elasticity (Micro) – Review

In this chapter, we will shift gears to the study of Microeconomics. Here we will look at the role of the consumer, revisit the laws of supply and demand, marginal utility, and the main ideas behind elasticity.

Let's begin by looking at a product market in equilibrium. The shaded region above the market price and under the demand curve is known as the **consumer surplus** *(see Diagram 62)*. When you buy a good at a price that is less than what you would have been willing to pay, you have a consumer surplus. If you are willing to buy a pair of socks for $3 and only pay $1, then you have a surplus of $2.

Diagram 62: Consumer Surplus & Producer Surplus

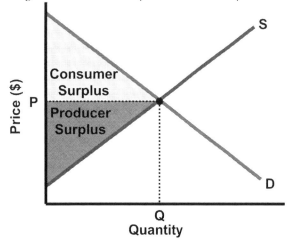

The *sum of consumer and producer surplus is maximized* when the free market is in equilibrium

It's also *allocatively efficient* and free of *deadweight loss*

Markets also create **producer surpluses**. This is the area below market price that is above the supply curve. This occurs when a seller receives a price that is greater than what he or she is willing to sell for, or the cost of production.

The demand curve is downward sloping for three reasons: **income effect** (a higher price means less purchasing power), **substitution effect** (at a higher price, buyers look for alternatives), and **diminishing marginal utility**.

Utility means satisfaction. As you consume more and more units of a good, such as a taco, your **total utility** will increase. However, the additional (marginal) utility will eventually fall; this is the law of diminishing marginal utility. What we are willing to pay for a good is based on our marginal utility that we receive from it.

Say you purchase 1 sandwich for lunch and observe an increase in total utility of 15. Because this is your first sandwich, marginal utility also increases by 15 *(see Diagram 63)*.

As you purchase more sandwiches for lunch, your total utility rises; however, the rate at which your total utility increases, eventually falls. When this rate of increase falls, the marginal utility diminishes. When you purchase that second sandwich, your marginal utility is 10, which is less than the additional satisfaction you experienced from the first sandwich. With each additional unit, marginal utility falls. Marginal utility is equal to the change in total utility divided by the change in quantity.

Diagram 63: Total Utility & Marginal Utility Example

Quantity	Total Utility	Marginal Utility
0	0	---
1	15	15
2	25	10
3	32	7
4	37	5
5	38	1

Marginal Utility $= \Delta TU / \Delta Q$

The marginal utility curve is the ***slope of the total utility curve***

Diminishing marginal utility occurs when total utility increases at a decreasing rate

Total utility is maximized when marginal utility is 0

Remember, marginal utility initially increases, then decreases. Diminishing marginal utility occurs when the total utility increases at a decreasing rate. Total utility is maximized when marginal utility is 0.

Let's say that you go to the mall to purchase shirts and pants with a certain amount of income in your wallet. You can use marginal utility to determine the optimal number of shirts and pants to purchase with your income. The **utility-maximizing formula** is as follows (*see Diagram 64*): Marginal Utility of Good X (shirts) divided by the Price (of shirts) is equal to the Marginal Utility of Good Y (pants) divided by the Price (of pants). If you need the marginal utility to decrease, then you would buy more. If you need the marginal utility to increase, then you would buy less.

Diagram 64: Consumer Equilibrium Equation

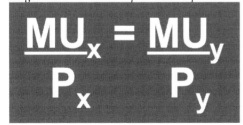

Buy more to reduce ***marginal utility*** and buy less to increase marginal utility

Let's move on to the **price elasticity of demand and supply**. Remember, elasticity measures responsiveness. Therefore, the formula for the price elasticity of demand is used to determine how responsive consumers are to changes in price.

An **elastic demand curve**, or one that is relatively flat, contains an elasticity that is greater than 1. Here, consumers are responsive to changes in price. These are goods that contain substitutes or goods that are not necessities. An **inelastic demand curve**, or a relatively steep demand curve, contains a price elasticity that is less than 1. Here, consumers are not very responsive to changes in price. These goods can be necessities or ones that use up a small portion of income.

When elasticity is equal to 1, we say it is **unit elastic**. If demand is **perfectly elastic**, the curve is a horizontal line, and if demand is **perfectly inelastic**, it is a vertical line.
As you move along a typical demand curve, demand is relatively price inelastic at lower prices and price elastic at higher prices.

When calculating the price elasticity of demand, you can ignore the negative sign. The formula is the percentage change in quantity demanded divided by the percentage change in price *(see Diagram 65)*.

Diagram 65: Price Elasticity of Demand Formula

$$E_d = \frac{\% \text{ Change in } Q_d}{\% \text{ Change in } P}$$

> **Elastic demand** ($E_d > 1$):
> Buyers are very responsive
> **Inelastic demand** ($E_d < 1$):
> Buyers are NOT very responsive

You can also estimate the elasticity of demand by using the **total revenue test**, a simple and useful tool. **Total revenue** is price multiplied by quantity *(see Diagram 66)*. If price rises and the total revenue falls, demand is most likely price elastic (greater than 1). If price rises and total revenue rises, demand is price inelastic (less than 1). And if price rises and total revenue stays the same, then demand is unit elastic (equal to 1).

Diagram 66: Total Revenue Formula

$$\text{Total Revenue} = P \times Q$$

> When price and **total revenue** move opposite ways, demand is **elastic**

The **price elasticity of supply** measures how responsive sellers are to changes in price. Because it takes time for sellers to adjust the quantity of economic resources employed to alter output, time is the most important determinant. Over the long run, supply is more price elastic than in the short run.

The formula for the price elasticity of supply is the percentage change in quantity supplied divided by the percentage change in price *(see Diagram 67)*.

Diagram 67: Price Elasticity of Supply Formula

$$E_s = \frac{\% \text{ Change in } Q_s}{\% \text{ Change in } P}$$

> **Supply is more elastic** in the long run than in the short run

Two other important types of elasticities are the income elasticity of demand and the cross elasticity of demand. We can use the **income elasticity of demand** to determine if a good is a normal good or an inferior good.

The formula to calculate the income elasticity of demand is percentage change in quantity demanded divided by the percentage change in income *(see Diagram 68)*. If the result is a positive number, it is a normal good (luxury goods are greater than 1, necessities are less than 1). If the result is negative, then it is an inferior good.

Diagram 68: Income Elasticity of Demand Formula

$$\text{Income } E_d = \frac{\% \text{ Change in } Q_d}{\% \text{ Change in Income}}$$

Normal (+) vs. ***Inferior*** (-)

The **cross elasticity of demand** is used to determine if two goods are substitutes, complements, or unrelated. The formula is the percentage change in quantity demanded of Good X divided by the percentage change in the price of Good Y *(see Diagram 69)*. If the number is positive, then the two goods are substitutes. If the number is negative, then the two goods are complements. If the number is close to zero, or zero, then the goods are unrelated.

Diagram 69: Cross Elasticity of Demand Formula

$$\text{Cross } E_d = \frac{\% \text{ Change in } Q_d \text{ of X}}{\% \text{ Change in Price of Y}}$$

Substitutes (+) vs. ***Compliments*** (-)

The worksheet and practice questions that follow will include the main ideas concerning elasticity and consumer behavior. You will also see questions on supply and demand, which we covered in NB1. Basic Concepts.

For the next two units, we will focus on the role of the producer in the market by summarizing the Costs of Production and the profit-maximizing behavior of firms in Product Markets.

Bonus Diagrams: Supply, Demand, & Government Actions

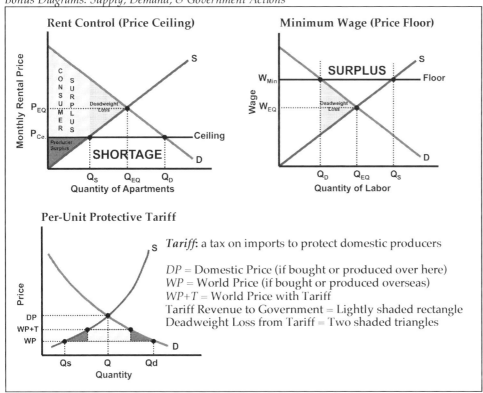

NB7. Utility & Elasticity (Micro) – Worksheet

Identify the following terms:

a. Consumer Surplus _____

b. Producer Surplus _____

c. Total Utility _____

d. Marginal Utility _____

e. Utility-Maximizing Formula _____

f. Elastic Demand Curve _____

g. Inelastic Demand Curve _____

h. Price Elasticity of Supply _____

i. Income Elasticity of Demand _____

j. Cross Elasticity of Demand _____

Summary questions:

1. How do you find the area of consumer surplus in a market? _____
   _____
   _____

2. How do you find the area of producer surplus in a market? _____
   _____
   _____

3. List and explain three reasons why demand is downward sloping. _____
   _____
   _____
   _____

4. What is the relationship between total utility and marginal utility? _____
   _____
   _____

5. How do you determine the optimum combination of consumer goods to purchase using
   the utility-maximizing formula? _____
   _____
   _____

6. What types of goods will most likely contain a price elasticity of demand greater than 1? Less than 1? _____

_____

_____

7. How do you use the total revenue test to estimate the price elasticity of demand? _____

_____

_____

8. How is the concept of time related to the price elasticity of supply? _____

_____

_____

9. How do you determine if a good is a normal good or inferior good using the income elasticity of demand formula? _____

_____

_____

10. How do you determine if two goods are substitutes, complements, or unrelated using the cross elasticity of demand formula? _____

_____

_____

NB7. Utility & Elasticity (Micro) – Practice

1. Which of the following shifts in demand and supply will lead to a decrease in equilibrium quantity and an indeterminate change in price?
   a. Demand shifts left; Supply shifts left
   b. Demand shifts right; Supply shifts right
   c. Demand shifts right; Supply shifts left
   d. Demand shifts left; Supply shifts right
   e. Cannot be determined based on the information given

2. The area beneath a good's demand curve and above the market price is known as the
   a. consumer surplus
   b. producer surplus
   c. inelastic region of the demand curve
   d. deadweight loss
   e. price ceiling

3. Suppose that the market for touchscreen MP3 players is in equilibrium at a market price of $199. The government decides to establish a ceiling price of $150 to support low-income consumers. Which of the following will occur?
   a. Demand will shift right
   b. Supply will shift right
   c. There will be a surplus of MP3 players
   d. There will be a shortage of MP3 players
   e. The market equilibrium will be unaffected

4. The producer surplus in a market can be described as the area
   a. to the right of the supply curve
   b. below the demand curve and above market price
   c. below the supply curve and above market price
   d. above the demand curve and below market price
   e. above the supply curve and below market price

5. The total utility curve initially _____ at an increasing rate then _____ at a decreasing rate.
   a. increases; increases
   b. increases; decreases
   c. decreases; decreases
   d. decreases; increases
   e. none of the above, total utility is constant

6. Suppose Liam drives to the stores in hopes of purchasing the ideal amount of books and DVD movies. If his marginal utility-price ratio for books is currently 5.7 at 7 units, and his marginal utility-price ratio for DVD movies is 6.9 at 4 units, how should Liam adjust his purchases?
    a. Buy more books and fewer DVD movies
    b. Buy fewer books and more DVD movies
    c. Buy more books and more DVD movies
    d. Buy fewer books and fewer DVD movies
    e. Liam should not alter his purchases

7. Suppose that the price elasticity of demand for dog collars is 0.74. If the price of dog collars _____ then total revenue will _____.
    a. increases; increase
    b. increases; decrease
    c. increases; stay the same
    d. decreases; increase
    e. decreases; stay the same

8. Which of the following is a determinant of the price elasticity of supply?
    a. Whether the product is a necessity
    b. The presence of substitute goods
    c. Time producers have to respond
    d. Percentage of income
    e. Whether the product is a luxury

9. Suppose that the percentage change in the quantity demanded of a good divided by a percentage change in income is positive. Which of the following must be true?
    a. This product is a necessity
    b. This demand is price elastic
    c. This product is a normal good
    d. This demand is price inelastic
    e. This product is an inferior good

10. Suppose the percentage change in quantity demanded of Good EN divided by the percentage change in price of Good OH is positive. Which of the following must be true?
    a. Good EN and Good OH are unrelated goods
    b. Good EN and Good OH are substitutes
    c. Good EN and Good OH are complements
    d. Good EN is inferior and Good OH is a normal good
    e. Good OH is inferior and Good EN is a normal good

NB7. Utility & Elasticity (Micro) – Practice Answers

1. A, Demand shifts left; Supply shifts left. When demand and supply shift left, the quantity will decrease. However, when demand shifts left, the price decreases. When supply shifts left, the price increases. Therefore, price is indeterminate and quantity decreases.

2. A, consumer surplus. The consumer surplus is the area above the market price and below the demand curve.

3. D, There will be a shortage of MP3 players. Because the ceiling (legal maximum) is set below free market equilibrium, there will be a shortage of touchscreen MP3 players.

4. E, above the supply curve and below market price. The area of producer surplus is the area above the supply curve and below the market price. The consumer surplus is the area below the demand curve and above the market price.

5. A, increases; increases. Total utility increases at an increasing rate then continues to increase at a decreasing rate. This is due to the law of diminishing marginal utility.

6. B, Buy fewer books and more DVD movies. In order for Liam to make his MU/P ratios equal, he must take into account the law of diminishing marginal utility. When he purchases fewer books, his MU/P for books will increase (5.7 increases). When he purchases more DVD movies, his MU/P for DVDs will decrease (6.9 decreases). Eventually the ratios will meet at Liam's consumer equilibrium.

7. A, increases; increase. Demand is price inelastic because it is less than 1. Total revenue will increase when the price rises (and total revenue will decrease when the price falls).

8. C, Time producers have to respond. When sellers have more time to respond to changes in the price, the supply tends to be price elastic. The other choices are determinants of the price elasticity of demand.

9. C, This product is a normal good. The purpose of the income elasticity of demand formula is to determine if a good is a normal good or an inferior good. It tells us how responsive consumers are to income changes. If it is positive, then the product is a normal good.

10. B, Good EN and Good OH are substitutes. The formula for the cross elasticity of demand is used to determine if two goods are substitutes or complements. If the cross elasticity is positive, then the goods are substitutes.

NB8. Costs of Production (Micro) – Review

In this chapter, we will focus on the role of the producer as we summarize the costs of production for a single firm.

First, let's get some basic definitions out of the way. **Economic costs** include explicit costs and implicit costs. **Explicit costs** are the monetary payments to resource suppliers for resources and **implicit costs** are the opportunity costs of self-owned economic resources.

To determine a firm's **economic profit**, simply calculate the total revenue then subtract total economic costs.

Because accountants do not take implicit costs into consideration, an **accounting profit** is equal to the total revenue minus explicit costs only.

In the **short run**, firms can increase or decrease output by relatively small quantities and cannot alter plant capacity. The time frame is simply too short. The firm can use any unused resources they own and hire some additional labor, but it will not build new plants or alter all of its resources.

In the **long run**, a firm's plant size can vary. The quantities of all resources can be changed. The time frame is also large enough for firms to enter or exit the market. Firms will enter the market in the long run when existing firms are earning short-run economic profits, and will exit the market when firms are taking short-run economic losses.

In NB7, we covered the concept of diminishing marginal utility for the consumer; this unit we will discuss **diminishing marginal returns** for the producer. As more units of labor (a variable resource) are added to existing fixed capital, the **total product** (output) increases. However, at some point the additional output, or **marginal product**, of the last worker hired will fall. This is known as diminishing marginal returns.

According to the table below *(see Diagram 70)*, when this particular firm hires its third worker, the marginal product decreases compared to the second worker hired. After the second worker is hired, the marginal product will diminish with each additional worker. The total product will continue to increase, however at a decreasing rate.

Diagram 70: Diminishing Marginal Returns Example

Labor	Total Product	Marginal Product	Average Product
0	0	---	---
1	22	22	22
2	46	24	23
3	65	19	21.67
4	81	16	20.25

> **Marginal Product** =
> $\Delta TP/\Delta Inputs$
>
> **Diminishing marginal returns**
> explains the shape of firms'
> product curves and cost curves

When you see the word *average*, simply divide the total by the quantity; in this case it's the quantity of workers. Total product divided by the number of laborers yields the **average product**. The average product curve will intersect the marginal product *(see Diagram 71)*, when average product is at its peak.

Diagram 71: Marginal & Average Product Curves

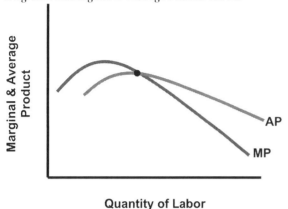

Diagity marginal returns — *Diminishing marginal returns* occurs when marginal product decreases and when **total product** increases at a decreasing rate

Average product is at its peak when it intersects marginal product

Short-run costs for a firm are either fixed or variable. **Fixed costs** must be paid whether the firm produces 100,000 units of output or 0 units of output. An example of a fixed cost is rent for land. **Variable costs** change with output, such as wages paid to hourly employees. **Total costs** are equal to the fixed cost plus variable cost.

Marginal costs are the most important costs for firms. These costs help determine a firm's profit-maximizing output. It is equal to the change in total cost divided by the change in output *(see Diagram 72)*. It is the change in cost from producing one additional unit of output. When marginal costs increase, the law of diminishing returns is at work.

Diagram 72: Marginal Cost Formula

$$\text{Marginal Cost} = \frac{\text{Change in TC}}{\text{Change in Q}}$$

A firm's **marginal costs** make up a firm's supply curve

Per-unit costs, or average costs, are important in determining profits and losses. The **average total cost** can be calculated by dividing total cost by quantity. The **average fixed cost** is equal to the total fixed cost divided by quantity. The **average variable cost** is equal to the total variable cost divided by quantity *(see Diagram 73)*.

Diagram 73: Per-Unit Cost Formulas

$$ATC = \frac{TC}{Q} \quad AFC = \frac{FC}{Q} \quad AVC = \frac{VC}{Q}$$

To calculate any **average**, simply divide totals by quantities

Graphically, the marginal cost curve initially decreases, bottoms out, and rises like a check mark *(see Diagram 74)*. The average total cost curve and the average variable cost curve are both u-shaped, and intersect the marginal cost curve at their minimums.

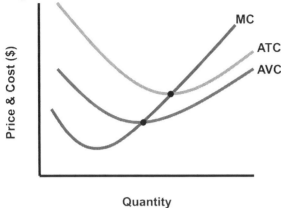

The **ATC** and **AVC** **curves** are at their minimum points when intersecting the **MC curve**

Marginal revenue is also important in determining the quantity of output that a firm produces. A profit-maximizing firm will produce where marginal revenue equals marginal cost. It is the additional revenue from producing one more unit of output *(see Diagram 75)*. Therefore, the marginal revenue is the change in total revenue divided by the change in output.

Diagram 75: Marginal Revenue Formula

MR = Change in Total Revenue
Change in Output

To calculate anything **marginal**, simply divide Δtotals by Δquantities

For a **perfectly competitive firm**, the marginal revenue curve is equal to the market price, the firm's demand curve, and the firm's average revenue curve. It is perfectly elastic or horizontal. As you can see on the graph below, the MR = MC intersection determines the optimal quantity the firm should produce *(see Diagram 76)*. This firm is earning a short-run economic profit because the price (P) exceeds the average total cost (ATC) curve at the MR = MC level of output. We will expand on this rule and model in NB9.

Diagram 76: Perfectly Competitive Firm Earning Short-Run Economic Profits, Price Exceeds ATC

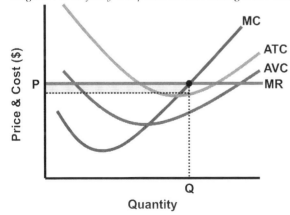

All firms will *maximize profit* where its marginal revenue equals its marginal cost

In the long run, the average total cost curve contains three parts: the first part is known as **economies of scale**; as output increases long-run average total costs fall *(see Diagram 77)*. Larger firms tend to enjoy this condition over a longer range of output than smaller firms. It serves as a **barrier to entry** in less competitive market structures like monopolies (1 seller) and oligopolies (2 to 4 sellers).

Diagram 77: Long-Run Average Total Cost Curve

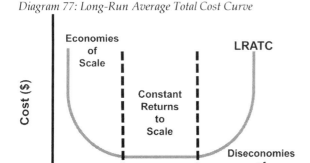

Economies of scale applies to long-run average total costs and serves as a ***barrier to entry*** in markets with a few powerful sellers

Minimum efficient scale is the first output point where LRATC reaches a minimum

The second part is known as **constant returns to scale**. As output increases, average total costs remain the same. The last part is called **diseconomies of scale**. In this region, as output increases the average total costs increase due to production complications.

Some firms operate in a constant-cost industry and some firms operate in an increasing-cost industry. In a **constant-cost industry**, many industries employ unspecialized economic resources. This means that if more firms enter one industry, resource costs will stay constant in the long run. In an **increasing-cost industry**, economic resources are specialized so more firms demanding specialized resources will increase the resource prices.

Be sure to complete the worksheet on the Costs of Production and to test your understanding with the practice questions. In NB9, we will summarize the four market structures and look at cost curves in more detail.

NB8. Costs of Production (Micro) – Worksheet

Identify the following terms:

    a. Economic Profit _____

    b. Accounting Profit _____

    c. Short Run _____

    d. Long Run _____

    e. Diminishing Marginal Returns _____

    f. Marginal Cost _____

    g. Average Total Cost _____

    h. Average Variable Cost _____

    i. Marginal Revenue _____

    j. Economies of Scale _____

Summary questions:

    1. What is the difference between an explicit cost and implicit cost? _____
_____
_____

    2. Why are accounting profits often higher than economic profits? _____
_____
_____

    3. Why might new firms enter an industry in the long run? _____
_____
_____

    4. Why might some firms exit an industry in the long run? _____
_____
_____

    5. What is the relationship between the average product curve and marginal product curve? _____
_____
_____

6. What is the difference between a fixed and variable cost? _____
   _____
   _____

7. How is the law of diminishing marginal returns reflected in the shape of the marginal cost curve? _____
   _____
   _____

8. What is the relationship between the marginal cost curve and average total cost curve?
   _____
   _____
   _____

9. What is the relationship between the marginal cost curve and average variable cost curve? _____
   _____
   _____

10. List and explain the three different parts of a long-run average total cost curve. Draw a long-run average total cost with each part labeled. _____
   _____
   _____
   _____
   _____
   _____

NB8. Costs of Production (Micro) – Practice

1. When marginal product falls and is greater than zero, total product
   a. increases at an increasing rate
   b. increases at a decreasing rate
   c. decreases at a decreasing rate
   d. decreases at an increasing rate
   e. remains constant

2. Suppose that a firm can produce:
   63 action figures when it hires 2 workers
   66 action figures when it hires 3 workers
   72 action figures when it hires 4 workers
   80 action figures when it hires 5 workers
   85 action figures when it hires 6 workers
   89 action figures when it hires 7 workers
   91 action figures when it hires 8 workers
   Diminishing marginal returns begins when this firm hires the
   a. 3rd worker
   b. 4th worker
   c. 5th worker
   d. 6th worker
   e. 7th worker

3. The chart below contains the short-run total costs for a broccoli farmer that sells its output in a perfectly competitive market. What is this firm's marginal cost of producing 6 units of output?

Output	Total Cost
0	45
1	51
2	64
3	78
4	98
5	120
6	147
7	175

   a. $13
   b. $20
   c. $22
   d. $27
   e. $28

4. An accounting profit is equal to
    a. total revenue - explicit and implicit costs
    b. total revenue - explicit costs
    c. total revenue - implicit costs
    d. total revenue - opportunity costs
    e. the total revenue

5. Suppose a firm's total costs are $2,000 at 5 units of output and average variable costs are $250 at 5 units of output. Calculate this firm's average fixed costs at 5 units of output.
    a. $150
    b. $650
    c. $850
    d. $1,750
    e. $2,250

6. Assume that total costs are $200 and total fixed costs are $50 when a firm produces 3 units of output. What is the average variable cost at 3 units?
    a. $25
    b. $50
    c. $75
    d. $150
    e. $200

7. The average total cost (ATC) curve is increasing
    a. when average variable costs are decreasing
    b. when marginal costs are less than ATC
    c. when average fixed costs are increasing
    d. when average variable costs exceed ATC
    e. after it intersects the marginal cost curve

8. Which of the following is true when this firm produces 70 units of output?

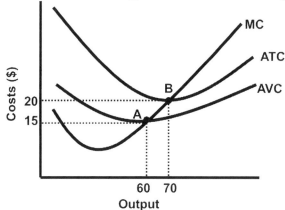

a. Marginal costs exceed average total costs
b. Average variable costs equal marginal costs
c. Average variable cost curve is decreasing
d. Marginal costs are $15
e. Average total cost curve is at a minimum

9. If total revenue increases at a constant rate then marginal revenue is_____.
a. perfectly elastic
b. perfectly inelastic
c. upward sloping
d. downward sloping
e. negative

10. When the long-run average total costs of a firm increase as output increases, it is experiencing
a. constant returns to scale
b. economies of scale
c. diseconomies of scale
d. diminishing returns to scale
e. diminishing utility to scale

NB8. Costs of Production (Micro) – Practice Answers

1. B, increases at a decreasing rate. As marginal product falls, total product continues to increase, but at a decreasing rate. This is known as diminishing marginal returns.

2. D, 6th worker. When this firm hires the 3rd worker, marginal product is 3 (66-63); when the firm hires the 4th worker, marginal product increases to 6 (72-66); at the 5th worker, marginal product increases to 8 (80-72). However, at the 6th worker marginal product falls to 5 (85-80). Marginal product falls because of diminishing marginal returns.

3. D, $27. The marginal cost of the sixth unit equals the change in total cost divided by change in output. Change in total cost is $27 (147-120) and change in output is 1 (6 - 5). 27 divided by 1 is equal to $27.

4. B, total revenue – explicit costs. Accounting profits do not take opportunity costs (implicit costs) into consideration. Only the explicit costs (payments to resource suppliers) are included.

5. A, $150. First calculate the firm's average total cost ($2,000/5 = $400). Next, subtract average variable costs from the average total costs to get the average fixed costs ($400 - $250 = $150).

6. B, $50. The variable cost is equal to the total cost ($200) minus the fixed cost ($50). The average variable cost is equal to the variable cost divided by output ($150/3 = $50).

7. E, after it intersects the marginal cost curve. After the average total cost curve intersects the marginal cost curve, ATC is increasing and marginal costs exceed ATC.

8. E, Average total cost curve is at a minimum. At 70 units of output, ATC intersects MC. Whenever the ATC curve intersects MC, ATC must be at a minimum.

9. A, perfectly elastic. If total revenue rises at a constant rate, the marginal revenue changes by the same amount at each level of output. Marginal revenue is perfectly elastic and equal to the product price for a perfectly competitive firm.

10. C, diseconomies of scale. When firms experience an increase in long run average total costs as output increases, it is called diseconomies of scale. It can be difficult to operate a firm efficiently when producing on a very large scale.

NB9. Product Markets (Micro) – Review

In this section, we will focus on the main ideas surrounding profit maximization in the four market structures.

Let's begin with a **perfectly competitive market** that contains many sellers producing identical products. The price of the product is determined by market supply and demand, not an individual firm. Here, the firm is a **price taker** because it *takes* the price from the market.

The marginal revenue for a perfectly competitive firm is equal to the price, the firm's perfectly elastic demand curve, and its average revenue curve. The firm can sell infinite units at this price so it would be foolish for a firm to lower its selling price.

To illustrate a **perfectly competitive firm earning a short-run economic profit**, label the vertical axis "price" and horizontal axis "quantity" or "output" *(see Diagram 78)*. Draw the MC curve first, followed by the ATC and AVC curves. The MR = MC intersection must occur above the ATC. The shaded area represents total economic profit. This profit is temporary and can only exist in the short run. In the long run, firms will enter the market causing price, or marginal revenue to fall. The firm will ultimately break even.

Diagram 78: Perfectly Competitive Firm Earning Short-Run Economic Profits, Price Exceeds ATC

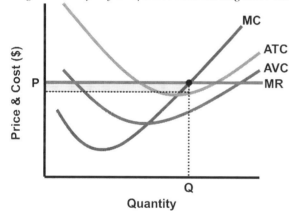

> ## Perfect Competition
> ***Short-run profit***: More firms will enter the market in the long run, which lowers the price (MR) and erases all of a firm's economic profit

To illustrate a **perfectly competitive firm minimizing economic losses** in the short-run, draw the cost curves in the same manner as the previous example. However this time, the MR = MC intersection must occur below the ATC *(see Diagram 79)*. The shaded area represents the firm's economic loss. In this example, the firm loses less money by producing at a loss rather than shutting down. In the long run, firms will exit the market, causing the price and marginal revenue to increase. Ultimately, the firm will break even.

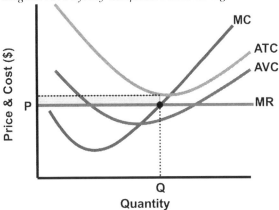

Diagram 79: Perfectly Competitive Firm Taking Short-Run Economic Loss, ATC Exceeds Price

Perfect Competition ***Short-run loss***: Weakest firms will exit the market in the long run, which raises the price (MR) and erases a firm's economic loss

When a firm cannot cover its variable costs and some of its fixed costs, it will **shut down** and still pay the fixed costs. In this example, the firm will lose less money by producing nothing. The MR = MC intersection must be below the AVC *(see Diagram 80)*. Firms will only produce above the minimum AVC; the **MC curve above the minimum AVC represents the firm's short-run supply curve**.

Diagram 80: Perfectly Competitive Firm: Short-Run Shut Down Case, AVC Exceeds Price

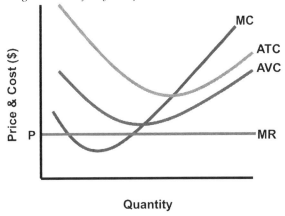

A firm will still pay fixed costs when it ***shuts down***

It loses less by producing zero units in the short run |

In the long run, the MR = MC intersection occurs at the minimum of the ATC curve *(see Diagram 81)*. This is known as **long-run equilibrium**, **zero economic profit**, **normal profit**, or **breakeven level of output**. Accounting profits are positive since accounting does not incorporate implicit costs.

Diagram 81: Perfectly Competitive Firm in Long Run Equilibrium, Price Equals ATC

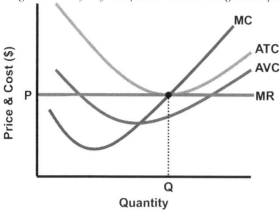

All perfectly competitive firms will **break even** in the long run and this is **highly efficient** (allocatively and productively)

In the long run, a perfectly competitive firm enjoys **productive efficiency** (produces at lowest cost) because it produces where the price equals minimum ATC. **Allocative efficiency** (optimal for society) also exists because the firm produces where price (demand) equals marginal cost (supply).

You should get in the habit of drawing the perfectly competitive market side-by-side with a perfectly competitive firm *(see Diagram 82)*. The market supply and demand determines the price, which is then sent over to the "price taking" firm. When shifts in the market occur, the firm's marginal revenue will shift up when the market price increases, or down when the market price decreases.

Diagram 82: Perfectly Competitive Market & PC Firm Earning Short-Run Economic Profits

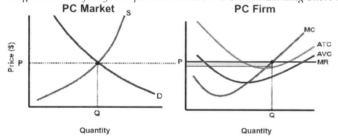

Draw your **perfectly competitive** markets and firms side-by-side

A **monopoly** exists when a market consists of one seller that offers a unique product. There are significant barriers to entry–such as economies of scale, legal patents, or government grants–making entry into the market impossible. The monopolist is a "price maker" and its demand curve is the industry's demand curve.

To draw a **monopolist earning an economic profit**, be sure to sketch a downward sloping demand curve above a downward sloping marginal revenue curve. The marginal cost curve and the average total cost curve are drawn the same way as a firm under perfect competition *(see Diagram 83)*.

The monopolist produces at the output level where MR = MC, however the price is on the demand curve. If the price is greater than the ATC, then the firm is earning an economic profit. The monopolist will sell fewer units at a higher price compared to a perfectly competitive firm, which creates an area of deadweight loss on the graph *(see Diagram 83a)*.

Diagram 83: Monopoly Earning Short-Run Economic Profits, Price Exceeds ATC

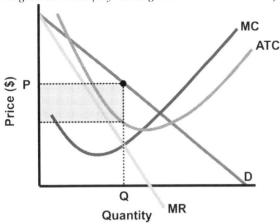

Demand, price, and ***average revenue*** are all same to a monopolist

A monopoly creates ***deadweight loss*** as it reduces consumer surplus

Diagram 83a: Monopoly Earning Short-Run Profits: Total Revenue, Total Costs, Consumer Surplus, & Deadweight Loss

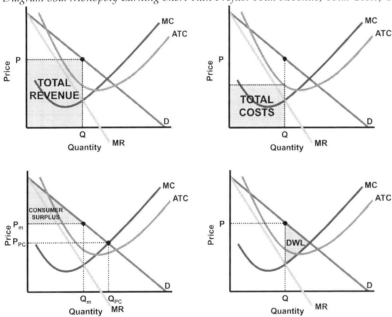

To show a **monopolist taking an economic loss**, the price on the demand curve must be below the ATC at the output level where MR = MC *(see Diagram 84)*. Please note that the demand curve is elastic when marginal revenue is positive, and inelastic when marginal revenue is negative. An unregulated monopolist will only produce in the elastic region.

Diagram 84: Monopoly Taking Short-Run Economic Loss, ATC Exceeds Price

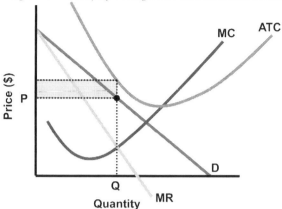

All firms experience an ***economic loss*** when demand is less than the average total cost curve at the MR=MC output point

A regulated monopoly might be forced to break even and sell at the **fair-return price**. This occurs when the price equals the ATC *(see Diagram 85)*. Notice that the price is not equal to MC so allocative efficiency is not achieved. Productive efficiency is also not achieved because the price is not equal to the minimum ATC.

Diagram 85: Regulated Monopoly: Fair-Return Price, Price Equals ATC

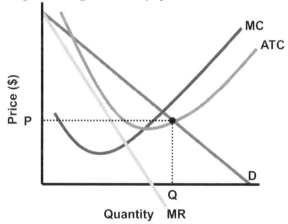

The ***fair-return price*** is not productively efficient

If a regulated monopolist produces the **socially optimal level of output,** it produces where price equals marginal cost *(see Diagram 86)*. This results in a lower price and greater quantity than the unregulated profit-maximizing monopolist. This point is allocatively efficient.

111

Diagram 86: Regulated Monopoly: Socially Optimal Price, Price Equals MC

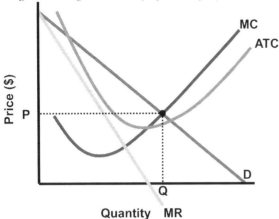

The ***socially optimal price*** is allocatively efficient because price equals marginal cost and this is where a perfectly competitive market would produce

If a monopolist practiced **perfect price discrimination**, consumer surplus would be eliminated *(see Diagram 86a)*. Since each buyer pays the highest price that he or she is willing to pay, the consumer surplus becomes part of the economic profit. The output level occurs where demand meets marginal cost.

Diagram 86a: Monopoly: Perfect Price Discrimination

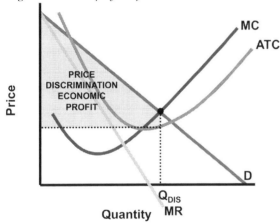

A ***perfectly price discriminating*** monopolist must prevent people from reselling its product

Consumer surplus becomes part of the economic profit

To illustrate a **monopolist maximizing total revenue**, the marginal revenue must equal 0 *(see Diagram 87)*. At this point, demand is unit elastic. The firm does not maximize its profit at this output level, only total revenue.

Diagram 87: Regulated Monopoly: Total revenue Maximization, MR Equals 0

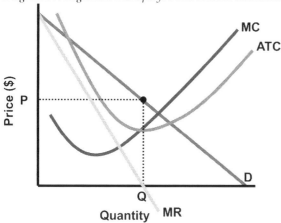

The ***total revenue maximization*** point occurs where demand is ***unit elastic*** because marginal revenue is 0

In the market structure of **monopolistic competition**, there are many firms selling similar products. Substitutes are available, however, the goods are not identical. There are elements of perfect competition and monopoly in this market structure.

The graph is similar to a monopoly in that the demand and marginal revenue curves are downward sloping. **In the long run, monopolistically competitive firms will break even**. The monopolistically competitive firm will also experience **excess capacity** because it produces at a point where ATC is still decreasing. The ATC will touch the demand curve, at the output level where MR = MC, but reach its minimum on the marginal cost curve *(see Diagram 88)*. This can be a little tricky to draw so be sure to practice.

Diagram 88: Monopolistically Competitive Firm in Long-Run Equilibrium, Price Equals ATC

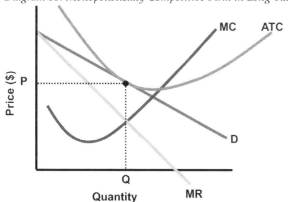

Monopolistically competitive graphs look the same as monopoly graphs except in the long run

In an **oligopoly**, there are a few firms (two to four) that produce similar or identical goods. One firm's strategy can determine another firm's profit. This is the market structure where we can apply **game theory**.

Suppose that there are two firms that sell skateboards: Totally Inc. and Awesome LLC *(see Diagram 89)*. The first number in each cell represents Totally Inc.'s possible payout and the second number is Awesome LLC.'s payout.

Diagram 89: Game Theory Example, Dominant Strategies Present

		Awesome LLC.	
		Strategy A	**Strategy B**
Totally Inc.	**Strategy A**	$2,305, $2,305	$2,350, $2,272
	Strategy B	$2,272, $2,350	$2,325, $2,325

> **If at least one firm has a *dominant strategy*…**

To determine the **dominant strategy**–the best outcome regardless of the other firm's strategy–look for the highest profits for each firm and ignore what the other firm's strategy might be. If Totally Inc. plays Strategy A, its payouts are $2,305 or $2,350, which are higher than its respective payouts of Strategy B, $2,272 and $2,325. Strategy A is Totally Inc.'s dominant strategy.

In this example, Awesome LLC. faces the same potential payouts as Totally Inc., so Strategy A is its dominant strategy as well. When both companies play their dominant strategies *(see Diagram 90)*, they arrive at the **Nash equilibrium**, which is highlighted below.

Diagram 90: Nash Equilibrium Example

		Awesome LLC.	
		Strategy A	**Strategy B**
Totally Inc.	**Strategy A**	$2,305, $2,305	$2,350, $2,272
	Strategy B	$2,272, $2,350	$2,325, $2,325

> **…then you can find the *Nash Equilibrium*…**

If the two firms **cooperate**, then they might agree to both play Strategy B where profits are higher than in the Nash equilibrium; however, the incentive to cheat arises because if one firm breaks the agreement to play Strategy B, it can earn an even higher profit. The other firm would lose profit unless it cheated too. This is known as the **prisoner's dilemma**.

Not every game contains dominant strategies for both firms. In the next example, two computer companies, Super Co and Duper Co, face the payouts shown below *(see Diagram 91)*. In this example, there are no dominant strategies and no Nash equilibrium.

Diagram 91: Game Theory Example, No Dominant Strategies

		Duper Co	
		Strategy A	**Strategy B**
Super Co	**Strategy A**	$2,450, 1,950	$5,950, $4,950
	Strategy B	$3,950, $4,450	$1,850, $2,750

> **…but a *game theory matrix* might not have any dominant strategies**

If Super Co plays strategy A, then its payout is either $2,450, which is less than the payout of $3,950 if it played strategy B, or it is $5,950, which is greater than $1,850. Duper Co's payouts work out in a similar manner. If Duper Co plays Strategy A, then its payout is either $1,950, which is less than $4,950 if it played strategy B, or it is $4,450, which is greater than $2,750.

You should be able to read a game theory matrix for the multiple-choice part of the exam as well as the free response. Game theory problems can be easy points, especially if all you have to do is identify payouts.

In NB10, we will review Factor Markets. First, go test your understanding of market structures by completing the worksheet and practice questions. Following the practice questions, you will find answers and explanations.

NB9. Product Markets (Micro) – Worksheet

Identify the following terms:

a. Perfectly Competitive Market _____

b. Short-Run Economic Profit _____

c. Short-Run Economic Loss _____

d. Long-Run Equilibrium _____

e. Productive Efficiency _____

f. Allocative Efficiency _____

g. Monopoly _____

h. Monopolistic Competition _____

i. Oligopoly _____

j. Dominant Strategy _____

Summary questions:

1. How do you illustrate a perfectly competitive firm earning short-run economic profits?
    _____
    _____
    _____
    _____
    _____

2. If perfectly competitive firms are earning short-run economic profits, what will occur in the long run? _____
    _____
    _____

3. If perfectly competitive firms are taking short-run economic losses, what will occur in the long run? _____
    _____
    _____

4. What is the relationship between a firm's marginal cost curve and its short-run supply curve? _____
    _____
    _____

5. How do you illustrate a monopoly earning short-run economic profits? _____

_____

_____

_____

_____

_____

6. How is the price elasticity of a monopolist's demand curve related to its marginal revenue curve? _____

_____

_____

7. How do you determine the level of output of a monopoly that is breaking even (fair-return price)? Socially optimal? Maximizing total revenue? _____

_____

_____

8. What are the differences between a monopoly and monopolistic competition? _____

_____

_____

9. How do you determine an oligopolist's dominant strategy using game theory? _____

_____

_____

10. How do you use game theory to determine the Nash equilibrium? _____

_____

_____

NB9. Product Markets (Micro) – Practice

1. Which of the following market structures contains firms that achieve both allocative and productive efficiency in the long run?
   a. Monopoly
   b. Monopolistic Competition
   c. Perfect Competition
   d. Oligopoly
   e. Perfect Competition and Monopolistic Competition

2. The chart below contains the short-run total costs for an asparagus farmer that sells its output in a perfectly competitive market. If the current market price is $18 per batch, what is the quantity that this firm will produce?

Output	Total Cost
0	45
1	51
2	64
3	78
4	98
5	120
6	147
7	175

   a. Between 2 and 3
   b. Between 3 and 4
   c. Between 4 and 5
   d. Between 6 and 7
   e. It will shut down and produce 0

3. Which of the following is true about firms within a perfectly competitive market?
   a. A PC firm's marginal revenue is equal to market price
   b. There are 3 or 4 firms in the PC market
   c. PC firms sell differentiated products
   d. PC firms can earn an economic profit in the long run
   e. PC firms charge a higher price than a monopolist

4. Which of the following is not true about a perfectly competitive firm in long-run equilibrium?
   a. The firm will break even
   b. Accounting profits are positive
   c. There are zero economic profits
   d. The firm is productively and allocatively efficient
   e. The firm is productively efficient, but not allocatively efficient

5. Firms maximize profits by producing at an output level where
    a. marginal utility is equal to price
    b. price is equal to total revenue
    c. marginal revenue is equal to marginal costs
    d. marginal revenue product is equal to marginal product
    e. total revenue is maximized

6. Suppose that a toy company is the only seller of dolls, and is currently enjoying economic profits. If the government passes a law that assigns a lump-sum subsidy to the company for the production of dolls, this firm's marginal costs will ____, price will ____, and profits will ____.
    a. increase; increase; decrease
    b. increase; increase; increase
    c. decrease; decrease; increase
    d. stay constant; stay constant; decrease
    e. stay constant; stay constant; increase

7. Assume that the graph below depicts the prices, costs, and revenues for an unregulated company that is the only seller of frozen pizza. At which set of points will this firm produce in the short run?

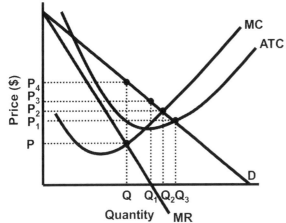

    a. P, Q
    b. P1, Q
    c. P2, Q2
    d. P4, Q
    e. P1, Q3

8. Which of the following is true concerning a monopolist and a perfectly competitive firm?
   I. A monopolist is a "price taker" and a perfectly competitive firm is a "price maker."
   II. A monopolist's marginal revenue is equal to the price.
   III. A monopolist and a perfectly competitive firm will produce at a level of output where marginal revenue equals marginal cost.
   IV. A perfect competitor's demand is equal to its marginal revenue.
   a. I only
   b. I and IV
   c. II only
   d. III only
   e. III and IV

9. Game theory and collusion are terms most closely associated with
   a. Perfect Competition
   b. Monopolistic Competition
   c. Oligopoly
   d. Monopoly
   e. Monopsony

10. Suppose that Slacker Inc. and Bored Corp. are the only two firms that sell mindless video games. The profits from choosing to create games like "This" or like "That" are represented in the matrix below. Slacker Inc.'s profits are given first in each cell.

		Bored Corp.	
		This	That
Slacker Inc.	This	$1,300, 800	$4,800, $3,800
	That	$2,800, $3,300	$700, $1,600

If Bored Corp. and Slacker Inc. cooperate, what will Bored Corp.'s profits be?
   a. $800
   b. $1,600
   c. $3,300
   d. $3,800
   e. $4,800

Free Response Practice

A: Suppose a perfectly competitive firm is currently earning short-run economic profits. Construct a **perfectly competitive market** and **perfectly competitive firm** side by side. Explain what will happen to the market and firm in the long run.

B: Construct a **monopoly** firm earning short-run economic profits. Identify and label the profit-maximizing quantity as "Q" and price as "P." Identify the location that the monopolist would sell at the fair-return price. Identify the socially optimal quantity.

NB9. Product Markets (Micro) – Practice Answers

1. C, Perfect Competition. Perfectly competitive firms break even in the long run. They produce at the lowest cost (productive efficiency: when price is equal to minimum of the average total cost curve) and the ideal amount for society (allocative efficiency: when price is equal to marginal cost).

2. B, Between 3 and 4. The marginal cost of producing 3 units is $14 ($78 - $64) and the marginal cost of 4 units is $20 ($98 - $78). Because this is a perfectly competitive firm, the price of $18 equals marginal revenue. The price of $18 lies between a quantity of 3 and 4.

3. A, A PC firm's marginal revenue is equal to market price. For a perfectly competitive firm, marginal revenue is horizontal at the market price (perfectly elastic). Total revenue increases at a constant rate. A monopolist sells its good at a higher price than a perfectly competitive firm.

4. E, The firm is productively efficient, but not allocatively efficient. Economic profits are zero, but accounting profits are positive. Accounting profits do not consider implicit costs. In the long run, a PC firm breaks even. Productive efficiency occurs when price equals minimum ATC, which happens in the long run, as does allocative efficiency (price = MC).

5. C, marginal revenue is equal to marginal costs. MR = MC represents the profit-maximizing point or loss-minimizing point for a firm. As long as the price is greater than AVC at this point, the firm will produce.

6. E, stay constant; stay constant; increase. A lump-sum subsidy will reduce average total costs while marginal costs, price, and output remain constant. A per-unit subsidy would change marginal costs, price, and output.

7. D, P4, Q. An unregulated monopolist will produce at the quantity where MR equals MC. The price is above that point on the demand curve (average revenue). Therefore, this monopolist will earn an economic profit at P4, Q.

8. E, III & IV. Firms in all market structures should produce where MR equals MC in order to maximize profits. For a perfectly competitive firm, the MR curve is the firm's demand curve, a perfectly elastic (horizontal) demand curve at the market price.

9. C, Oligopoly. The game theory matrix and the concept of collusion are terms most commonly associated with an oligopoly. In an oligopoly there are very few sellers of similar products.

10. D, $3,800. Both firms can make the most profit if Slacker Inc. chooses "This" and Bored Corp. chooses "That." If they agree to play these strategies then Bored Corp. receives $3,800 profit, while Slacker Inc. makes $4,800.

Free Response Solution

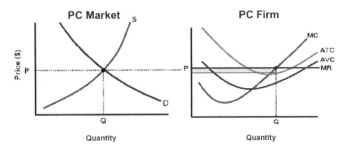

A: The graphs above show a perfectly competitive market alongside a firm earning short-run economic profits. In the long run, more firms will enter the market which will shift market supply to the right. The price will decrease and the firm's marginal revenue will shift down to meet the minimum average total cost curve. The firm achieves both allocative and productive efficiency in the long run.

B: The monopolist will produce where MR=MC and will earn economic profits because the price exceeds ATC at this level of output.

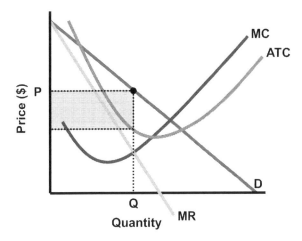

The fair-return price occurs when the price (demand) is equal to ATC. The socially optimal price occurs when price is equal to MC.

NB10. Factor Markets (Micro) – Review

In this section, we will focus on the main ideas concerning the functions of factor markets.

The first thing you need to understand is that households and businesses exchange roles in factor markets as households supply resources and firms demand resources. The demand for an economic resource is derived from the product market. This concept is known as **derived demand**.

In NB8, we covered the concept of diminishing marginal returns, which demonstrates that the marginal product falls as additional workers are hired by a firm. This rule also applies to a firm's marginal revenue product curve, which is another name for the firm's resource demand curve.

The **marginal revenue product** is the additional revenue that a firm receives from hiring an additional unit of a resource such as a worker *(see Diagram 92)*. It is the change in total revenue divided by the change in quantity of resources, or the marginal product multiplied by the marginal revenue. You should be able to work out the marginal product and marginal revenue product if given a chart containing the quantity of resources and total product data.

Diagram 92: Marginal Revenue Product Formula

$$MRP = \frac{\text{Change in Total Revenue}}{\text{Change in Quantity of Resources}}$$

> **Marginal revenue product** is the demand for a resource

The **determinants of marginal revenue product** include: demand for the product, productivity, and the prices of other resources. Changes in any of these areas will result in a shift of the marginal revenue product curve.

The **marginal factor cost** is the additional cost of hiring an additional resource. A profit-maximizing firm will hire workers, or any resource, up to the point where the marginal revenue product equals the marginal factor cost. A firm that hires labor in a perfectly competitive labor market will find that the wage equals the marginal factor cost as well as the firm's labor supply curve.

A **perfectly competitive labor market** determines the wage rate, which is a resource payment. The firm becomes the **"wage taker."** Therefore, the firm's MFC curve, or supply curve, is perfectly elastic. The **MRP = MFC** point represents the profit-maximizing quantity of workers that this firm should hire *(see Diagram 93)*. Here, it would not increase its profit by hiring an additional worker.

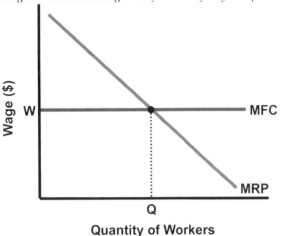

> **Marginal factor cost**
> is the supply for a
> resource in a perfectly
> competitive market

If a single firm experiences an **increase in labor productivity**, then the firm will want to hire more of those productive laborers *(see Diagram 94)*. If the marginal product curve shifts upward, then the marginal revenue product curve will shift right.

Diagram 94: Firm Hiring from a Perfectly Competitive Labor Market, MRP Shifts Right

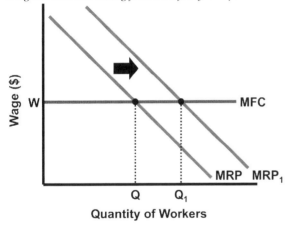

> An increase in
> productivity shifts the
> **marginal product** curve
> and the marginal revenue
> product curve of an
> economic resource

You should be comfortable drawing the perfectly competitive labor market and firm side-by-side, similar to what we did in NB9 with perfectly competitive product markets *(see Diagram 95)*. Supply and demand determines the equilibrium wage in the labor market and the firm pays that market wage to its workers.

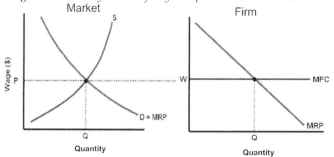

Draw **perfectly competitive** resource markets and firms side-by-side

If there is a rightward shift of supply in the labor market, then the wage will fall *(see Diagram 96)*. For the firm, the MFC = Wage curve will shift down resulting in a greater quantity of workers hired.

Diagram 96: Market Supply Shifts Right Causing the Firm's Wage (MFC) Decreases

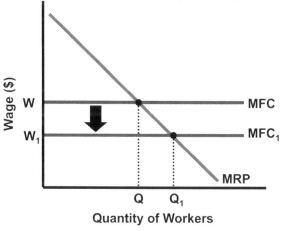

The firm is a **wage taker** when it hires from a perfectly competitive labor market

A **monopsony** exists when only one firm buys labor or any economic resource. Since there is only one firm that demands labor, the firm becomes the **"wage maker."** For a monopsony, the MFC curve lies above the labor supply curve *(see Diagram 97)*. The firm will hire the quantity of workers where marginal revenue product equals the marginal factor cost, but will pay the lowest possible wage, the point on the labor supply curve. Due to this "hiring power," fewer workers will be hired and wages will be lower, compared to a perfectly competitive labor market.

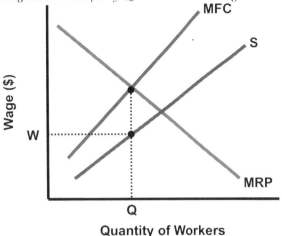

Diagram 97: Monopsony: Q = MRP = MFC, Wage on the Supply Curve

A ***monopsony*** is a wage maker that will employ fewer workers at a lower wage than competitive labor markets

When a firm utilizes two types of resources, like labor and capital, there are two important equations to know. The purpose of these equations is to determine the ideal quantity of factors to employ.

One equation is known as the **least-cost rule**. This can be calculated by dividing the marginal product of a resource (labor) by its price, and setting it equal to the marginal product of another resource (capital) divided by that resource's price *(see Diagram 98)*. If you need the marginal product to decrease, then the firm should employ more of the resource due to the law of diminishing marginal returns. If you need the marginal product to increase, then employ fewer units of the resource.

Diagram 98: Least-Cost Rule of Resources Formula

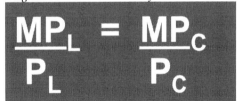

$$\frac{MP_L}{P_L} = \frac{MP_C}{P_C}$$

Employ more resources to reduce ***marginal product*** and employ fewer resources to increase marginal product

The other equation is known as the **profit-maximizing rule**. The marginal revenue product of a resource (labor) divided by the price of the resource should equal the marginal revenue product of the other resource (capital) divided by that resource's price *(see Diagram 99)*. It is similar to the least-cost rule, except the profit-maximizing ratios must equal 1. If you need the marginal revenue product to decrease, then the firm should employ more of the resource due to the law of diminishing marginal returns. If you need the marginal revenue product to increase, then employ fewer units of the resource.

$$\frac{MRP_L}{P_L} = \frac{MRP_C}{P_C} = 1$$

Employ more resources to reduce **MRP** and employ fewer resources to increase MRP

Let's wrap up our review of Factor Markets with a quick note on **unions**. A major goal of all unions is to increase wages. To obtain higher wages, a union can attempt to boost product demand, which increases the demand for their labor; or a union can reduce the supply of workers by requiring employee licensing; or a union can organize in a way where the members set a minimum wage rate above the market equilibrium (similar to a price floor or minimum wage) to control the supply curve of labor.

Even though we focused on labor markets in this chapter, you can also apply these principles to the other types of economic resources.

In NB11, we will complete our review of Microeconomics by discussing the role of the government. First, test your skills by completing the worksheet and practice questions on Factor Markets. Following the practice questions and graphing practice, you will find the answers and explanations.

NB10. Factor Markets (Micro) – Worksheet

Identify the following terms:

a. Derived Demand _____

b. Diminishing Marginal Returns _____

c. Marginal Revenue Product _____

d. Marginal Factor Cost _____

e. Perfectly Competitive Labor Market _____

f. Monopsony _____

g. Least-Cost Rule (Combination of Resources) _____

h. Profit-Maximizing Rule (Combination Resources) _____

Summary questions:

1. What are the roles of businesses and households in resource (factor) markets? _____
   _____
   _____

2. What are the determinants of the marginal revenue product curve? _____
   _____
   _____

3. How is the wage rate related to marginal factor cost for a firm that hires its labor from a perfectly competitive labor market? _____
   _____
   _____

4. Why is a firm that hires labor from a perfectly competitive labor market called a "wage taker"? _____
   _____
   _____

5. How do you determine the profit-maximizing quantity of workers for a firm to hire? ___
   _____
   _____

6. How do you illustrate a firm that hires its workers from a perfectly competitive labor market? _____

_____

_____

_____

_____

_____

7. How does an increase in labor productivity experienced by a single firm affect the wage rate and quantity of workers hired? _____

_____

_____

8. How does the wage rate and number of workers employed by a monopsony compare with the wage rate and number of workers employed in a perfectly competitive labor market? _____

_____

_____

9. How do you illustrate a monopsony hiring its profit-maximizing quantity of labor? ____

_____

_____

10. What are three ways in which unions attempt to increase the wages of employees? _____

_____

_____

_____

NB10. Factor Markets (Micro) – Practice

1. Suppose that an increase in demand for automobiles increases the demand for auto mechanics. This is an example of
   a. derived demand
   b. marginal revenue costs
   c. marginal factor costs
   d. a monopsony
   e. an oligopoly

2. How can one calculate the marginal revenue product?
   a. Marginal physical product x Price of a resource
   b. Change in total revenue/Change in output
   c. Change in total revenue/Change in inputs
   d. Marginal physical product x Change in output
   e. Marginal revenue x average revenue

3. Which of the following does the demand (D) and supply (S) of labor represent for this firm?

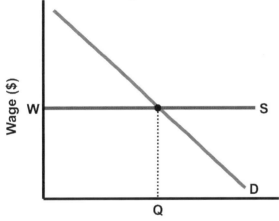

Quantity of Workers

   a. D = MC; S = MR
   b. D = MFC; S = MRP
   c. D = TR; S = MR
   d. D = MRP; S = MFC
   e. D = MU; S = TU

4. Which of the following will not cause an increase in labor demand?
   a. Increased product supply
   b. Increased product price
   c. Increased productivity
   d. Increased product demand
   e. Increased price of a substitute resource

5. Which of the following could have caused the shift in MRP shown below?

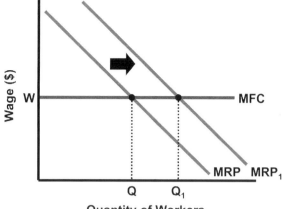

Quantity of Workers
   a. An increase in the wage rate
   b. A decrease in the wage rate
   c. An increase in this firm's labor productivity
   d. A decrease in this firm's labor productivity
   e. A decrease in product price

6. If you were to compare a perfectly competitive labor market and a firm that has a monopsony, you would find that a monopsony hires
   a. more workers at a lower wage
   b. more workers at a higher wage
   c. fewer workers at a higher wage
   d. fewer workers at a lower wage
   e. the same amount of workers at the same wage

7. Which of the following points correctly illustrates the wage that this firm will pay its workers and the quantity of workers it will hire?

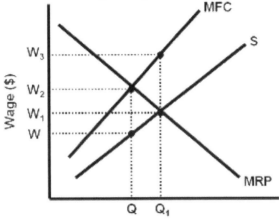

a. W, Q
b. W1, Q1
c. W2, Q
d. W3, Q
e. W3, Q1

8. An effective minimum wage is an example of a _____, which must be _____ the equilibrium wage and leads to a _____ of workers in the labor market.
a. price ceiling; below; surplus
b. price ceiling; above; shortage
c. price ceiling; above; surplus
d. price floor; below; shortage
e. price floor; above; surplus

9. Suppose a firm employs two types of economic resources known as "peh" and "kew." Which of the following formulas best illustrates the profit-maximizing combination of resources?
a. $MPP_{peh}/MPP_{kew} = MRP_{peh}/MRP_{kew}$
b. $MRP_{peh}/P_{peh} = MRP_{kew}/P_{kew} = 1$
c. $MPP_{peh}/P_{peh} = MPP_{kew}/P_{kew}$
d. $MPP_{peh}/1 = MPP_{kew}/1$
e. $1/MPP_{peh} = MPP_{kew}/1$

10. Suppose there are only two resources needed in the production of magic wands: capital and labor. Piper's Wizardry is currently minimizing costs by producing at a level where the marginal product of capital is 50 wands per hour while capital is $10 per hour. If the cost of labor is $6 per hour, what is the marginal product of labor per hour?
    a. 5 wands
    b. 6 wands
    c. 25 wands
    d. 30 wands
    e. 50 wands

NB10. Factor Markets (Micro) – Practice Answers

1. A, derived demand. The demand for mechanics is derived from the product market, which is the car market in this case. More cars will mean an increase in demand for mechanics to perform oil changes, inspections, and repairs.

2. C, Change in total revenue/Change in inputs. MRP equals the change in total revenue divided by a change in variable inputs (or) MRP = Marginal physical product x Marginal revenue.

3. D, D = MRP; S = MFC. A firm's resource demand curve is known as the marginal revenue product (MRP) and the supply is the marginal factor cost (MFC). In this case, the MFC is equal to the wage, which is constant.

4. A, Increased product supply. Increased product supply (rightward shift of supply) results in a lower price in the market and would not increase resource demand. However, an increase in product supply could be a result of increased productivity. An increase in the price of a substitute resource, such as capital, will increase labor demand.

5. C, An increase in this firm's labor productivity. An increase in this firm's productivity of labor will increase MRP and therefore its demand for workers. The wage in the labor market remains unchanged, as does this firm's wage.

6. D, fewer workers at a lower wage. A monopsony is the "wage maker" that pays the lowest wages possible and hires fewer workers. A firm that hires workers in a perfectly competitive labor market is a "wage taker." Because many firms are hiring, more workers will be hired in a PC labor market.

7. A, W, Q. This is a monopsony. This firm is the only one that hires a certain type of labor. Therefore, it is a "wage maker" and will pay the lowest wage that it can. MRP = MFC tells us the quantity of workers (Q). The lowest wage that it can pay this quantity of workers is on the supply curve (W).

8. E, price floor; above; surplus. A minimum wage is a type of price floor. This means that it must be above the equilibrium wage rate to be effective. It will lead to a surplus of workers because more people will seek employment at a higher wage. However, firms are not willing to hire all of these workers at the higher minimum wage.

9. B, $MRP_{peh}/P_{peh} = MRP_{kew}/P_{kew} = 1$. The profit-maximizing combination of resources can be found with this formula: $MRP_x/P_x = MRP_y/P_y = 1$. Do not confuse this with the least-cost rule $MP_x/P_x = MP_y/P_y$.

10. D, 30 wands. Use the cost-minimizing rule for combining resources. The MP of capital divided by the price of capital equals the MP of labor divided by the price labor ($50/10 = x/6$). When you solve for x, you will get a marginal product of labor of 30 magical wands.

NB11. The Government (Micro) – Review

In this summary, we will review the role of government within the economy and certain markets. For one, the government provides **public goods**: goods that people share and that people cannot be excluded from consuming. The two characteristics of a public good are **shared consumption** (non-rivalrous) and **non-exclusion**. Two classic examples are local police departments and the military.

It is believed that the private market will not provide the socially optimal quantity of a public good, which is why the government gets involved. These should be goods that produce positive externalities and make society better off. The **free rider problem** is an issue that arises when discussing public goods. For example, people that do not pay taxes may enjoy the benefits of police or military protection. It is very difficult to prevent someone that didn't pay for a public good from enjoying the benefits.

The government imposes taxes so it can carry out its functions. The two main principles behind taxation are the **ability-to-pay principle** (a tax based on what you can afford) and the **benefits-received principle** (a tax based on what you receive).

One type of tax is a **progressive tax** such as a personal income tax. This means that a greater proportion of income is paid by those with higher levels of income than those with lower levels of income. A **regressive tax** is the opposite, which has people with lower income levels paying a greater proportion of income. Finally, a **proportional tax** sees the proportion remain constant at all income levels.

Whenever the government implements a **per-unit tax** on a good, you should recall that the market supply shifts left *(see Diagram 100)*. The result is that the buyer and seller both pay a portion of the tax. "P" represents the price before the tax and "P1" represents the price that the buyer pays. The top rectangle is the portion of the tax that the buyer pays and the bottom rectangle is what the seller pays. Together, these two rectangles represent the total tax revenue that goes to the government.

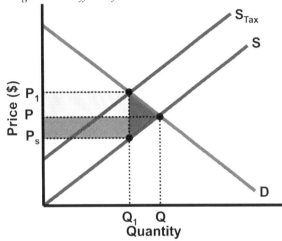

Diagram 100: Effects of a Per-Unit Tax

Price does not rise by the full amount of a *per-unit tax* since buyers and sellers both pay part of the tax

"Ps" is the price that the seller receives. If a $5 per-unit tax occurs, the price paid by the consumer will not increase by the full $5 because the tax is shared. The shaded triangle represents the loss of efficiency, which is known as **deadweight loss**. It is inefficient for the government to tax unless the government is correcting a negative externality.

In the graph above, the buyers and sellers pay fairly equal portions of the tax. However, that is not always the case. When demand is more elastic than supply, the seller will pay a greater portion, and when demand is more inelastic than supply, the buyer will pay a greater portion of the tax (*see Diagram 100a*).

Diagram 100a: Per-Unit Tax Examples with Elasticity

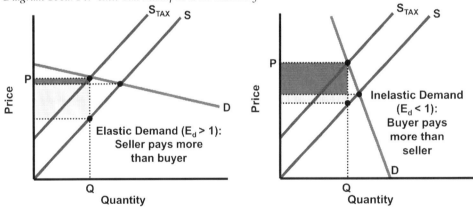

Sometimes private markets create **externalities** where third parties are affected by a market transaction. The marginal benefit curve represents the demand curve and the marginal cost curve represents supply.

Negative Externality: If the production of a good results in pollution, there are social harms, or **marginal social costs** (*see Diagram 101*). The market price is "P" however the marginal social costs are greater. There is an **overallocation of resources**. This means that the marginal social costs exceed the marginal social benefits. The market quantity is "Q," but should only be "Q1." The shaded triangle represents the deadweight loss. Remember, externalities are inefficient.

Diagram 101: Negative Externality: Marginal Social Costs Exceed Marginal Social Benefits

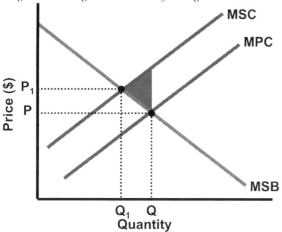

Correct a ***negative externality*** with a per-unit tax to reduce or eliminate the deadweight loss

The **government can correct this negative externality** by taxing the production of this good. This would shift the supply curve left toward "Q1" where the MSC curve sits.

Positive Externality: When the production of a good generates positive externalities, there are higher marginal social benefits than marginal social costs (*see Diagram 102*). The market price and quantity is "P" and "Q" respectively, however the social benefits are higher. The equilibrium quantity should be at "Q1." The deadweight loss is represented by the shaded triangle.

Diagram 102: Positive Externality: Marginal Social Benefit Exceeds Marginal Social Cost

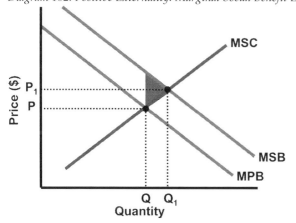

Correct a ***positive externality*** with a per-unit subsidy to reduce or eliminate the deadweight loss

The **government can correct this positive externality**, or **underallocation of resources**, by subsidizing the buyer or seller, which will increase the quantity toward "Q1."

The government also attempts to limit the degree of **income inequality** within an economy. This can be achieved with taxes (like the personal income tax system) and transfer payments

137

(such as welfare). The government can shift the **Lorenz curve** (*see Diagram 103*), which illustrates the degree of income inequality, inward or toward perfect income equality.

Diagram 103: Lorenz Curve

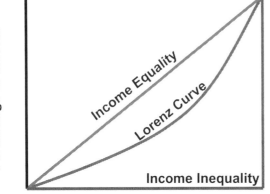

Percentage of Households

A ***Gini coefficient*** closer to 1 indicates greater income inequality than a Gini coefficient that is closer to 0

Those are some of the main concepts regarding the Government and its economic functions. This chapter concludes our review of Microeconomics. You should also check out *No Bull Review's Top Ten Guide* for even more important concepts, formulas, and graphs.

Before exam day, be sure that you can draw all of the graphs, as well as their variations, discussed throughout the No Bull Unit Reviews and practice questions. I recommend drawing the following 5 graphs and models over and over again before exam day.

No Bull Tip – Master these 5 Microeconomics graphs!
1. **Supply & Demand**
2. **Perfectly Competitive Market & Firm Side-By-Side**
3. **Monopoly Firm**
4. **Perfectly Competitive Labor (Resource) Market & Firm Side-By-Side**
5. **Positive & Negative Externalities**

You should also complete recent part two free-response questions that are available on the Internet for free.

Go complete the worksheet and practice questions on The Government, then take the No Bull Microeconomics Exam. Good luck!

NB11. The Government (Micro) – Worksheet

Identify the following terms:

a. Public Good _____

b. Ability-to-Pay Principle _____

c. Benefits-Received Principle _____

d. Progressive Tax _____

e. Regressive Tax _____

f. Proportional Tax _____

g. Deadweight Loss _____

h. Negative Externality _____

i. Positive Externality _____

j. Lorenz Curve _____

Summary questions:

1. What are the two characteristics of a public good? _____
   _____
   _____

2. What is the free rider problem associated with public goods? _____
   _____
   _____

3. What are the effects of a per-unit tax on market price and quantity? Which group(s) pay(s) for the tax? _____
   _____
   _____

4. Where is the deadweight loss on a graph that shows the effects of a per-unit tax? Draw and explain. _____
   _____
   _____
   _____
   _____

5. How do you illustrate a market with negative externalities? _____

_____

_____

_____

_____

_____

6. Where is the deadweight loss on a graph that shows negative externalities? _____

_____

_____

7. How can the government correct a negative externality? _____

_____

_____

8. How do you illustrate a market with positive externalities? _____

_____

_____

_____

_____

_____

9. Where is the deadweight loss on a graph that shows positive externalities? _____

_____

_____

10. How can the government correct a positive externality? _____

_____

_____

NB11. The Government (Micro) – Practice

1. Two characteristics that a public good must possess are
    a. shared consumption and exclusion
    b. shared consumption and non-exclusion
    c. non-shared consumption and non-exclusion
    d. non-shared consumption and exclusion
    e. rivalrous and exclusion

2. If every household paid an income tax that is equal to 23% of their income, then the income tax is _____.   If households with lower incomes are taxed a greater percentage of income than high income households, then the income tax is _____. If households with higher incomes are taxed a greater percentage of income than lower income households, then the income tax is _____.
    a. proportional; progressive; regressive
    b. regressive; proportional; progressive
    c. proportional; regressive; progressive
    d. regressive; progressive; proportional
    e. progressive; regressive; proportional

3. Suppose that the supply of Good Dee is price inelastic and demand is relatively price elastic. If the government imposes a per-unit tax on Good Dee then
    a. the consumer and producer equally share the tax
    b. producers will increase production
    c. the consumer will pay a greater portion of the tax
    d. the producer will pay a greater portion of the tax
    e. Cannot be determined based on the information given

4. The graph below illustrates the externalities associated with the sale of flying broomsticks.

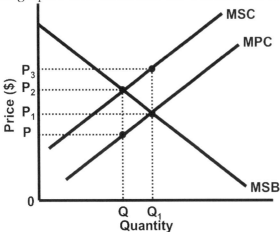

I. There are spillover benefits from the sale of flying broomsticks.
II. There is an overallocation of resources to the production of flying broomsticks.
III. The government can correct this spillover by providing subsidies to buyers.
IV. The government can correct this spillover by providing subsidies to producers.
    a. II only
    b. III and IV
    c. I and III
    d. I, III, and IV
    e. I, II, III, and IV

5. Suppose that society gets too little of Good Ess for too low of a price. Which of the following should occur to correct this problem?
    a. Demand for Good Ess should shift left
    b. Supply of Good Ess should shift left
    c. Demand and Supply should both decrease
    d. A per-unit tax on the sale of Good Ess
    e. A per-unit subsidy on the production of Good Ess

6. The graph below shows the market for jelly donuts after a per-unit tax. Which shaded region represents the portion of the tax paid by the buyers?

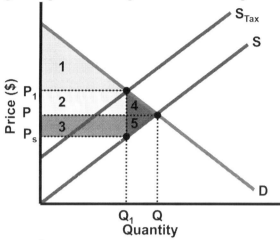

   a. 1
   b. 2
   c. 3
   d. 4
   e. 5

7. The graph below shows the market for scented candles after a per-unit tax. Which price represents the price received by the sellers?

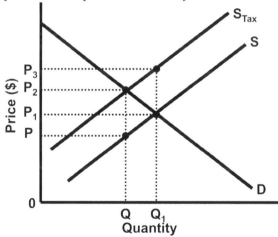

   a. P
   b. P1
   c. P2
   d. P3
   e. 0

8. Suppose that the supply and demand of Good ZEE have similar price elasticities. If the government imposes a sales tax on Good ZEE then
    a. the consumer will pay a greater portion of the tax
    b. the producer will pay a greater portion of the tax
    c. the government will pay a greater portion of the tax
    d. the supply will shift right
    e. the consumer and producer shares the tax equally

9. In mixed market economic systems, why does the government enact laws that limit the formation of trusts?
    a. To lower the marginal costs for firms
    b. To raise total revenue for firms
    c. To promote competition in the market place
    d. To grant fewer firms the exclusive right to produce goods
    e. To increase the market the price of goods

10. Which of the following government actions can best shift the Lorenz curve inward?
    a. Implementing a regressive income tax system
    b. Implementing a proportional income tax system
    c. Implementing a progressive income tax system
    d. Reducing estate taxes
    e. Reducing taxes on the top 1% of income earners

NB11. The Government (Micro) – Practice Answers

1. B, shared consumption and non-exclusion. Two classic characteristics of a public good are shared consumption (non-rivalrous) and non-exclusion.

2. C, proportional; regressive; progressive. The average tax rate stays the same regardless of income for proportional taxes. Progressive taxes are ones in which higher income people are taxed a greater percentage of their income.

3. D, the producer will pay a greater portion of the tax. When supply is relatively inelastic, the producer feels the burden of the tax more than the consumer.

4. A, II only. The social costs are greater than private costs. This means that negative externalities exist. There are too many broomsticks so there is an overallocation of resources. The government can correct this externality by taxing the production of flying broomsticks.

5. E, A per-unit subsidy on the production of Good Ess. If society is getting too little of a good at too low of a price, then positive externalities are present. The government can subsidize production to increase the quantity. This would shift supply to the right.

6. B, 2. The region marked with a 2 represents the portion of the tax that the buyer pays. The region marked with a 3 represents the portion of the tax that the seller pays. Together, regions 2 and 3 make up the total tax revenue.

7. A, P. The price received by the sellers after the tax is P. P1 is the price before the tax and P2 is the price that the buyers pay.

8. E, the consumer and producer shares the tax equally. When the price elasticities of supply and demand are equal, both consumers and producers share the tax burden equally.

9. C, to promote competition in the market place. One of the key functions of the government is to maintain competition in markets. Anti-trust legislation limits the creation of monopolies and achieves this objective.

10. C, Implementing a progressive income tax system. A progressive tax system is best for reducing overall income inequality, which means the Lorenz curve can shift inward from this government action.

No Bull Exam – Macroeconomics

Part I: Answer all 60 multiple-choice questions by choosing the letter of the best answer.

1. Suppose that in a given year government spending is $22 billion, imports are $12 billion, transfer payments are $2 billion, consumption is $62 billion, exports are $9 billion, and the gross domestic product is $100 billion. Based on the above data, what is this economy's gross investment?
   a. $13 billion
   b. $17 billion
   c. $19 billion
   d. $21 billion
   e. $22 billion

2. An increase in which of the following will cause the shift in aggregate demand shown below?

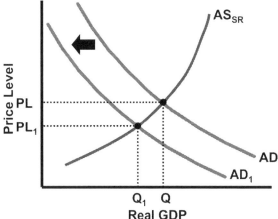

   a. Federal funds rate
   b. Disposable income
   c. Money supply
   d. Government spending
   e. Open market purchases of bonds by the Fed

3. If the firms within a given economy cut their costs of production, aggregate supply shifts _____, prices _____, and real GDP _____.
   a. right; increase; increases
   b. left; decrease; decreases
   c. right; decrease; increases
   d. left; increase; decreases
   e. left; increase; increases

4. If the supply of beeswax increases and the demand for beeswax increases simultaneously, then what will happen to the market price and quantity of beeswax?
   a. Price increases; Quantity indeterminate
   b. Price indeterminate; Quantity increases
   c. Price indeterminate; Quantity decrease
   d. Price decrease; Quantity indeterminate
   e. Price increases; Quantity increases

5. Suppose that the government cuts personal income taxes by $25 billion and the marginal propensity to save is 0.25. As a result of this tax cut, output will increase by
   a. $8 billion
   b. $25 billion
   c. $33 billion
   d. $75 billion
   e. $100 billion

6. Assume that the Federal Reserve employs open market operations when the economy is operating in the Keynesian range of the aggregate supply curve. As a result of the monetary policy shown below, how will output and unemployment change?

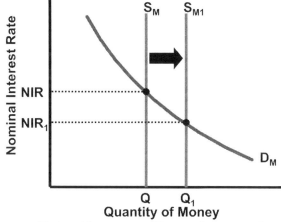

   a. Output increases; Unemployment does not change
   b. Output decreases; Unemployment increases
   c. Output does not change; Unemployment decreases
   d. Output increases; Unemployment decreases
   e. Output increases; Unemployment increases

7. Which of the following would not be counted toward a nation's gross domestic product in the current year?
   I. A 2007 automobile sold at a used car dealership in April.
   II. The commission that the salesman earned from selling the 2007 automobile.
   III. Stocks purchased with household income in July.
   IV. Rent paid for a 2-bedroom apartment in February.
   a. I only
   b. I and III
   c. II and IV
   d. I, III, and IV
   e. I, II, III, and IV

8. Which of the following groups will not benefit from unanticipated inflation?
   I. Creditors
   II. Those paying a 30-year fixed mortgage
   III. Those living on a fixed income
   IV. Debtors
   a. I and III
   b. II and IV
   c. III and IV
   d. I, III, and IV
   e. I, II, III, and IV

9. If the economy is struggling with a recession, a suitable fiscal policy would include an increase in
   a. purchases of securities by the Fed
   b. government spending
   c. taxes
   d. the discount rate
   e. exports

10. Which flow represents goods and services in the circular flow diagram shown below?

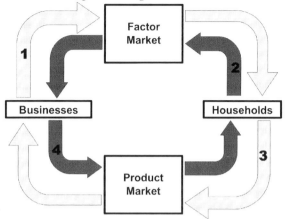

   a. 1
   b. 2
   c. 3
   d. 4
   e. All of the above

11. Assume ukuleles are inferior goods and that consumer income declines. Which of the following is true within the market for ukuleles?
   a. Demand for ukuleles will shift right
   b. Market price of ukuleles will fall
   c. Supply of ukuleles will decrease
   d. Supply of ukuleles will increase
   e. Quantity of ukuleles supplied will decrease

12. Assume that the gross domestic product has reached its minimum and the unemployment rate is at its peak. In which section of the business cycle is this economy currently operating?
   a. Expansionary
   b. Peak
   c. Contractionary
   d. Trough
   e. Recovery

13. The market for wire hangers is currently in equilibrium and a wire hanger is an inferior good. If consumer income rises at the same time that the costs of producing wire hangers falls, then what will happen to the market equilibrium?
   a. Price indeterminate; Quantity increases
   b. Price indeterminate; Quantity decreases
   c. Price decreases; Quantity indeterminate
   d. Price decreases; Quantity decreases
   e. Price increases; Quantity indeterminate

14. Which of the following will shift right as a result of the leftward shift shown below?

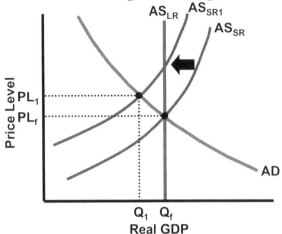

    a. Aggregate demand curve
    b. Long-run Phillips curve
    c. Money supply curve
    d. Short-run Phillips curve
    e. Demand curve for loanable funds

15. The type of inflation that is brought on by a negative supply shock and is associated with a leftward shift of the aggregate supply curve is known as
    a. cost-pull inflation
    b. cost-push inflation
    c. demand-push inflation
    d. demand-pull inflation
    e. stagnation inflation

16. Assume that the supply of alphorns, a normal good, increases and the demand for alphorns increases simultaneously. All of the following could have caused any of these shifts to occur except
    a. an increase in subsidies to alphorn producers
    b. an increase in the price of a substitute for alphorns
    c. an increase in the number of sellers of alphorns
    d. an increase in the price of a complement to alphorns
    e. an improvement in the productivity of creating alphorns

17. Suppose that the economy is currently functioning beneath its full-employment level of output. A decrease in which of the following would be a suitable monetary policy?
    a. Personal income taxes
    b. International value of the dollar
    c. Required reserve ratio
    d. Buying government bonds
    e. Gross investment

18. According to the circular flow model, _____ earn revenue in the product market and _____ sell resources in the factor market.
    a. businesses; households
    b. households; businesses
    c. businesses; businesses
    d. households; households
    e. households; the government

19. All of the following are examples of economic resources except
    a. farmland in Saskatchewan
    b. a treasury note worth $5,000
    c. a robotic arm in a Tuscaloosa, Alabama automobile plant
    d. yak wool used in the production of sweaters
    e. a power plant in Springfield

20. If the Federal Reserve executes an expansionary monetary policy, interest rates _____, investment and consumer expenditures _____, aggregate demand shifts _____, output and prices _____, and unemployment _____.
    a. decrease; decrease; right; decrease; falls
    b. decrease; increase; right; increase; falls
    c. decrease; increase; right; decrease; falls
    d. increase; decrease; left; decrease; rises
    e. increase; increase; right; increase; falls

21. "Extra! Extra! Read All About It! Today, the government issued a statement that eating one cube of Gouda cheese can cause a severe, possibly deadly, stomach virus." Based on the hypothetical (and ludicrous) headline above, what could happen to the market price and quantity of Gouda cheese?
    a. Price increases; Quantity increases
    b. Price decreases; Quantity decreases
    c. Price increases; Quantity decreases
    d. Price decreases; Quantity indeterminate
    e. Price indeterminate; Quantity decreases

22. Which of the following will most likely cause a decrease in the unemployment rate when the economy is in recession?
    a. An increase in personal income taxes
    b. An increase in the federal funds interest rate
    c. An increase in imports
    d. An increase in consumer expenditures
    e. An increase in the international value of the dollar

23. Which of the following will cause movement from point C to point B?

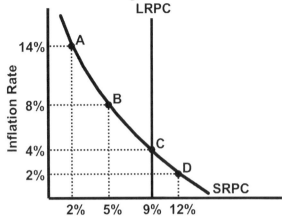

a. Rightward shift of aggregate demand
b. Rightward shift of short-run aggregate supply
c. Leftward shift of money supply
d. Leftward shift of short-run aggregate supply
e. Leftward shift of aggregate demand

24. Assume that real interest rates are 12% and the expected rate of inflation is 10%. What is the nominal interest rate?
    a. -2%
    b. 2%
    c. 10%
    d. 12%
    e. 22%

25. Suppose that there is an independent change in investment spending of $100 million and that the marginal propensity to consume is 0.75. As a result of this change in investment spending, output will increase by
    a. $20 million
    b. $75 million
    c. $133 million
    d. $200 million
    e. $400 million

26. An increase in which of the following could have caused the shift shown below?

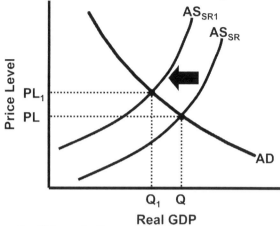

a. Price of oil
b. International value of the dollar
c. Imports from China
d. Federal funds rate
e. Personal income taxes

27. The Federal Open Market Committee targets the federal funds rate. How can the F.O.M.C. cause an increase in the federal funds rate?
a. Selling government securities
b. Decreasing the discount rate
c. Buying government bonds
d. Decreasing the required reserve ratio
e. Decreasing levels of spending

28. The government decides that it would like to support low-income consumers in the market for computers since technology has become a vital ingredient to fostering learning and improving education. Which of the following courses of action would the government take to meet its goal?
a. Establish a price floor below equilibrium price in the computer market
b. Establish a price ceiling below equilibrium price in the computer market
c. Impose an excise tax on computer sales
d. Impose a value-added tax on the stages of computer production and distribution
e. Decrease subsidies to producers of computers

29. Refer to the information in the previous question: What is the likely effect of the government's course of action on the market for computers?
a. The supply of computers will shift to the right
b. The demand for computers will increase
c. There will be a surplus of computers
d. There will be a shortage of computers
e. Nothing will happen, market remains in its original equilibrium

30. Fill in the blanks based on the graph shown below: _____ in personal income taxes while maintaining current levels of government spending will shift demand to the right causing private borrowing to _____.

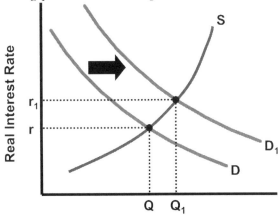

a. A decrease; decrease
b. An increase; increase
c. A decrease; increase
d. An increase; decrease
e. A decrease; remain constant

31. If the Federal Reserve makes a decision to increase the money supply, how will aggregate demand (AD) and bond prices change?
    a. AD increases; Bond prices increase
    b. AD increases; Bond prices decrease
    c. AD increases; Bond prices stay the same
    d. AD decreases; Bond prices increase
    e. AD decreases; Bond prices decrease

32. How will the supply of US dollars to the foreign exchange market and the international value of the dollar change as a result of the increase in real GDP shown below?

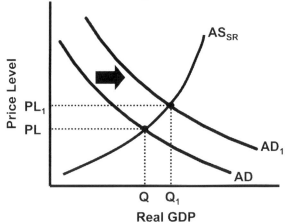

   a. Supply increases; Dollar appreciates
   b. Supply increases; Dollar depreciates
   c. Supply decreases; Dollar depreciates
   d. Supply decreases; Dollar appreciates
   e. Supply stays the same; Dollar depreciates

33. Suppose the economy is operating at a level of output that is below full employment and there are no changes in fiscal and monetary policies. What will happen to real GDP and the price level as the economy moves toward long-run equilibrium?
   a. Real GDP decrease; Price level decrease
   b. Real GDP decrease; Price level increase
   c. Real GDP increase; Price level increase
   d. Real GDP increase; Price level decrease
   e. Real GDP no change; Price level no change

34. Assume that Toll Booth Johnny collects tolls on a major bridge. One day Johnny receives notification that he will be replaced by a computerized toll collection system and released from employment. Which of the following types of unemployment will Johnny experience?
   a. Frictional
   b. Seasonal
   c. Structural
   d. Cyclical
   e. Transitional

35. Assume that Chasquack can produce one computer using 20 units of resources and one bicycle using 10 units of resources. Innduck can produce one computer using 10 units of resources and one bicycle using 30 units of resources. Which country has the absolute advantage in computer production and bicycle production?
    a. Chasquack for computers, Innduck for bicycles
    b. Chasquack for computers and bicycles
    c. Innduck for computers, Chasquack for bicycles
    d. Innduck for computers and bicycles
    e. Cannot be determined from the information given

36. Refer to the information given in the previous question concerning Chasquack and Innduck: Will 1 bicycle for 2 computers be acceptable terms of trade for Chasquack and Innduck?
    a. Yes for Chasquack, Yes for Innduck
    b. Yes for Chasquack, No for Innduck
    c. No for Chasquack, No for Innduck
    d. No for Chasquack, Yes for Innduck
    e. Cannot be determined from the information given

37. If the economy is operating at the full-employment level of output, how might a tight monetary policy affect the unemployment rate and bond prices?
    a. unemployment increases; bond prices increase
    b. unemployment increases; bond prices decrease
    c. unemployment decreases; bond prices decrease
    d. unemployment decreases; bond prices increase
    e. unemployment decreases; bond prices remain constant

38. The graph below shows an economy operating at the full-employment level of output. If there is a decrease in gross investment, how will price level and real GDP change?

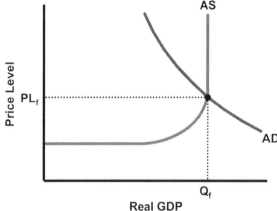

    a. Price level increase; Real GDP increase
    b. Price level increase; Real GDP decrease
    c. Price level stays constant; Real GDP decrease
    d. Price level decrease; Real GDP decrease
    e. Price level decrease; Real GDP stays constant

39. Which of the following is not included in M1?
    a. Checkable deposits
    b. Coin
    c. Traveler's checks
    d. Currency
    e. Institutional money market funds

40. Suppose that the Federal Reserve enacts a tight monetary policy at the same time that the government enacts an expansionary fiscal policy. How will the international value of US dollars and the level of US investment spending be affected?
    a. Dollar appreciates; Investment increases
    b. Dollar appreciates; Investment decreases
    c. Dollar depreciates; Investment increases
    d. Dollar depreciates; Investment decreases
    e. Dollar depreciates; Investment is indeterminate

41. Suppose that 2004 has a GDP deflator of 100 and a real GDP of $14 trillion. Which of the following must be true?
    I. Cost-push inflation exists.
    II. 2004 is the base year.
    III. Cyclical unemployment exists.
    IV. Nominal GDP is $14 trillion.
    a. I and II
    b. I and III
    c. II and III
    d. II and IV
    e. I, II, and IV

42. Long-run economic growth will most likely occur when there is an increase in
    a. consumption
    b. interest rates
    c. education and training
    d. transfer payments
    e. net exports

43. Which of the following will not cause the shift in supply shown below?

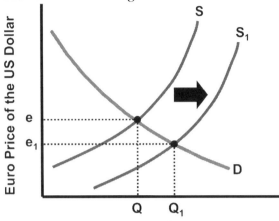

a. Contractionary US fiscal policy
b. Increased popularity of German cars
c. Increased income in the eurozone
d. Expansionary US monetary policy
e. Increased real income in the US

44. Refer to the foreign exchange market graph in the previous question: Which of the following is true concerning the change shown in the foreign exchange market?
   a. The quantity of dollars supplied decreased
   b. The quantity of dollars demanded decreased
   c. US dollars are more expensive for Europeans
   d. The euro appreciated in value relative to the US dollar
   e. The demand for euros decreased

45. If the nominal interest rate is 9% and the rate of inflation is 7%, what is the real interest rate?
   a. -2%
   b. 2%
   c. 7%
   d. 9%
   e. 16%

46. Suppose that the required reserve ratio is 10% and banks lend out all of their excess reserves. Ross the Boss earned $1,500 cash yesterday and decides to deposit $1,000 of his earnings in his checking account while stashing the other $500 under his mattress. What is the maximum change in the money supply from Ross' $1,000 deposit?
   a. $1,000
   b. $1,500
   c. $9,000
   d. $10,000
   e. $13,500

47. Assume that the economy is currently operating in the Keynesian range of the aggregate supply curve. If there is a decrease in personal income taxes, what will happen to disposable income, price level, and output?
    a. Disposable Income increases; Price Level increases; Output increases
    b. Disposable Income decreases; Price Level decreases; Output decreases
    c. Disposable Income increases; Price Level increases; Output decreases
    d. Disposable Income decreases; Price Level decreases; Output increases
    e. Disposable Income increases; Price Level decreases; Output increases

48. All of the following will cause interest rates to increase except
    a. a rightward shift of the money demand curve
    b. an increase in personal income taxes
    c. a leftward shift of the supply of loanable funds
    d. an increase in government spending
    e. selling of government bonds by the Fed

49. The interest rate that the Federal Reserve charges banks for short-term loans is known as the
    a. federal funds rate
    b. annual percentage rate
    c. prime rate
    d. discount rate
    e. treasury rate

50. Which graph is most closely related to the one shown below?

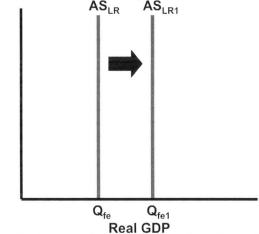

    a. Rightward shift of the supply of loanable funds
    b. Rightward shift of the production possibilities curve
    c. Rightward shift of aggregate demand in the short run
    d. Rightward shift of money demand curve
    e. Rightward shift of the short-run Phillip's curve

51. Suppose that velocity increased and the nominal gross domestic product stayed the same. Which of the following should be true?
    a. Money stock increased
    b. Money stock decreased
    c. Money stock remained the same
    d. Prices increased
    e. Prices decreased

52. Suppose that the Chinese spend $20 million on US treasury bonds. This action can be considered a _____ in the US _____ account.
    a. credit; current
    b. credit; capital
    c. credit; federal funds
    d. debit; capital
    e. debit; current

53. Assume that the economy is currently operating at its full-employment level of output. In the short run, demand pull-inflation could result from
    a. a tightened money supply
    b. an increase in the required reserve ratio
    c. an increase in imports
    d. lower interest rates
    e. decreased spending by the government

54. Which of the following could have caused the shift in demand shown below?

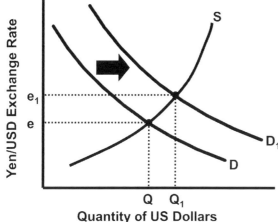

    a. Lower relative price level in Japan
    b. A tight US monetary policy
    c. A contractionary US fiscal policy
    d. Increased real income in the US
    e. Increased tastes for Japanese goods

55. Suppose that the government pursues a policy to shrink the federal budget deficit. Which of the following will not decrease?
    a. Aggregate demand
    b. Interest rates
    c. Unemployment
    d. International value of the currency
    e. Price level

56. In an economy, the marginal propensity to consume is 0.6 and the government is operating with a balanced budget. If the government increases spending and taxes by $250 million, what will happen to output?
    a. Increase by $250 million
    b. Increase by $375 million
    c. Increase by $416.7 million
    d. Increase by $500 million
    e. Increase by $625 million

57. In recent decades, the Federal Reserve has been no stranger to criticism. Rational expectations theory suggests that when the Fed increases the money supply to boost employment, inflation expectations _____ and real output _____.
    a. rise; increases
    b. rise; decreases
    c. rise; stays the same
    d. fall; increases
    e. fall; decreases

58. Suppose that the exchange rate between the Vietnamese dong and the Bangladeshi taka are 270 dong per 1 taka. If the exchange rate changed to 260 dong per 1 taka then the Vietnamese dong _____, causing Vietnam to import _____ goods from Bangladesh. Vietnam's current account balance will move toward a _____.
    a. appreciated; more; deficit
    b. appreciated; fewer; deficit
    c. depreciated; more; deficit
    d. depreciated; fewer; surplus
    e. depreciated; more; surplus

59. Under which of the following conditions should the Federal Reserve enact a policy that results in a leftward shift of the money supply?
    a. Full employment equilibrium
    b. Low prices
    c. Low exports
    d. High unemployment
    e. High rate of inflation

60. Suppose that households decrease their savings, which affects the equilibrium real interest rate in the loanable funds market. How will the change in the real interest rate affect the international demand of the dollar and net exports?
    a. demand for the dollar increases; Net exports increase
    b. demand for the dollar decreases; Net exports decrease
    c. demand for the dollar decreases; Net exports increase
    d. demand for the dollar increases; Net exports stay constant
    e. demand for the dollar increases; Net exports decrease

Part II: Answer all 3 free-response questions and include all necessary graphs and lines of economic reasoning.

1. Presently, an economy is experiencing a 2% rate of inflation and 11.6% rate of unemployment. Draw an AD/AS graph for this economy that displays these economic conditions.
    a. Label the current level of output "Q" and the price level "PL."
    b. Indicate the full-employment level of output with the label "Q_f"
    c. What is a prudent fiscal policy action for this economy? Explain.
    d. What is a sensible open market operation for this economy? Explain.
    e. If no policies are employed, explain how this economy will return to full employment in the long run.

2. Draw a money market graph in equilibrium.
    a. If the demand for money increases, what will happen to the nominal interest rate? Show on your graph.
    b. How can the Fed counteract this change in the nominal interest rate? Explain.
    c. Suppose that the Fed buys $30 million of treasuries when the reserve ratio is 25%. How much will the total money supply increase from the Fed's action?
    d. If you deposit $200 into your checking account, by how much can the money supply potentially increase?

3. Suppose that an economy's short-run aggregate supply curve decreased.
    a. Draw a short-run Phillips curve that shows the effects of the change in short-run aggregate supply.
    b. Suppose that due to the change in real income caused by the shift in short-run aggregate supply, Americans supply fewer dollars to the foreign exchange market. What will happen to net exports in the US?
    c. How is this change in net exports related to the short-run Phillips curve?

_____
_____
_____
_____
_____
_____

No Bull Exam – Macroeconomics Answers

1. C, $19 billion. Consumption (C) + Gross Investment (Ig) + Government Spending (G) + Net Exports (Xn) = Gross Domestic Product (62 + Ig + 22 + 9 - 12 = 100). Ignore transfer payments because they do not reflect production.

2. A, Federal funds rate. An increase in the federal funds rate (bank-to-bank interest rate) makes borrowing more difficult. The private sector (C + Ig) will spend less and aggregate demand will shift left.

3. C, right; decrease; increases. When the aggregate supply curve shifts right, the price level falls and real GDP increases. AS shifts right when the resource costs decrease.

4. B, Price indeterminate; Quantity increases. If supply of beeswax shifts right, price decreases and quantity increases. If demand for beeswax shifts right, price increases and quantity increases. Therefore, price is indeterminate because it can increase, decrease, or remain constant. The market quantity will definitely increase.

5. D, $75 billion. The tax multiplier is equal to MPC divided by the MPS (0.75/0.25 = 3). The change in taxes multiplied by the tax multiplier will yield the change in output ($25 billion x 3 = $75 billion).

6. D, Output increases; Unemployment decreases. Due to the lower nominal interest rate, households and businesses will borrow more money for consumer and investment expenditures. Aggregate demand will increase, causing output to increase and unemployment to decrease.

7. B, I and III. The 2007 car was counted in 2007's GDP; however the commission of the salesman is income for the current year. Stocks are purely financial transactions and do not count toward GDP. Rent is a service so it counts as part of the GDP even though the building was built in an earlier year.

8. A, I and III. People paying loans will pay them off with cheaper dollars, which hurts the creditor (benefits the debtor). People on a fixed income will receive money with less purchasing power. The creditors and people on fixed incomes will lose.

9. B, government spending. Fiscal policy includes changes in government spending and/or taxes. An increase in government spending is an expansionary policy appropriate during a recession.

10. D, 4. Businesses provide households with goods and services through the product market.

11. A, Demand for ukuleles will shift right. If income falls, the demand for inferior goods will increase (Price increases; Quantity increases).

12. D, Trough. The trough is the bottoming out of economic activity. It is the worst that the recession has to offer.

13. C, Price decreases; Quantity indeterminate. When the costs of producing wire hangers fall,

supply shifts to the right (P falls, Q rises). Since a wire hanger is an inferior good, demand shifts left when income increases (P falls, Q falls). After both shifts, the quantity can increase, decrease, or stay the same.

14. D, Short-run Phillips curve. When short-run aggregate supply shifts left, the short-run Phillips curve will shift right. This is due to the simultaneous increase in both prices and unemployment.

15. B, cost-push inflation. A supply shock, such as a rapid increase in the price of an economic resource like oil, will lead to a leftward shift of aggregate supply and cost-push inflation.

16. D, an increase in the price of a complement to alphorns. If the price of a complementary good for alphorns rises, then the demand for alphorns would decrease.

17. C, Required reserve ratio. If the economy is in recession, an appropriate monetary policy is to reduce the reserve requirement so banks have more money in excess reserves to lend out. This can be a powerful tool because a change in the reserve ratio will change the amount of excess reserves in the banking system and the money multiplier.

18. A, businesses; households. Businesses earn revenue from the products they sell to households. Households supply the land, labor, capital, and entrepreneurship to the businesses in the factor (resource) market.

19. B, a treasury note worth $5,000. Financial assets, securities, and money are not considered economic resources. Capital consists of tools, machines, and factories.

20. B, decrease; increase; right; increase; falls. Easy monetary policies result in lower interest rates, which encourage private spending (C + Ig). This will increase aggregate demand, RGDP, and the price level. The unemployment rate will decrease.

21. B, Price decreases; Quantity decreases. If something is known to be bad for one's health, demand for that product tends to decline. Consumer tastes and preferences would cause the demand to shift left (Price decreases; Quantity decreases).

22. D, An increase in consumer expenditures. An increase in consumer spending, the largest component of GDP, will reduce unemployment. An increase in the value of the dollar will mean more imports and fewer exports, which reduces GDP.

23. A, Rightward shift of aggregate demand. When aggregate demand shifts right, the price level rises (inflation increases) and unemployment decreases. Shifts in AD cause point-to-point movement along the short-run Phillips curve.

24. E, 22%. Use the Fisher Equation: Nominal Interest Rate = Real Interest Rate + Inflation Rate (N.I.R. = 12% + 10%).

25. E, $400 million. If the marginal propensity to consume (MPC) is 0.75, the marginal propensity to save (MPS) must be 0.25. The spending multiplier is 1 divided by the MPS (1/0.25 = 4). The change in output is the initial change in spending multiplied by the multiplier ($100 x 4 = $400).

26. A, Price of oil. Because oil is such an important economic resource, an increase in the price of oil will cause the short-run aggregate supply curve to shift left. The price level will rise and output will decrease.

27. A, Selling government securities. When the Fed sells government bonds, money exits the banking system causing the federal funds rate to increase. The discount rate and reserve ratio are not part of the Fed's open market operations.

28. B, Establish a price ceiling below equilibrium price in the computer market. An effective price control for this situation is a price ceiling, which must be set below the free market's equilibrium price. Computer prices will be artificially low and result in a shortage of computers. The quantity demanded will exceed the quantity supplied.

29. D, There will be a shortage of computers. A price ceiling that is set below the equilibrium price will cause a shortage. The quantity demanded will exceed the quantity supplied.

30. A, A decrease; decrease. A decrease in taxes will reduce the government's tax revenue. This means that the government will borrow money to maintain spending. Demand for loanable funds increases causing a higher interest rate. Private spending and borrowing will decrease due to the higher real interest rate. This is known as crowding out.

31. A, AD increases; Bond prices increase. If there is an increase in the money supply, nominal interest rates will fall. Private spending, aggregate demand, and output will increase. Because people have more money, they will purchase more financial assets. There will be greater demand for bonds, which will raise bond prices. Bond prices and interest rates are inversely related.

32. B, Supply increases; Dollar depreciates. If there is an increase in real US income, then Americans will have more dollars to supply to the foreign exchange market. The increase in supply of dollars means that the dollar will depreciate.

33. D, Real GDP increase; Price level decrease. If the economy is in a recession and policymakers do nothing, eventually long-run equilibrium will be restored. Businesses will demand fewer laborers and wages will fall. The cost of production has decreased so the short-run aggregate supply will shift right to meet the full-employment output level. The price level falls, real GDP increases, and unemployment falls.

34. C, Structural. When technology replaces workers, structural unemployment exists. Johnny will have to retrain or move to find employment.

35. C, Innduck for computers, Chasquack for bicycles. Absolute advantage means: which economy can produce more, faster, or with the least number of resources? It takes Innduck less resource units to produce a computer (10<20) and Chasquack less resource units to produce a bicycle (10<30).

36. A, Yes for Chasquack, Yes for Innduck. Chasquack will specialize in bicycle production because it has a lower relative opportunity cost (1/2<3) and Innduck will specialize in computers (1/3<2). According to the terms of trade, Chasquack would import 2 computers for each bicycle it exports, while Innduck would receive 1/2 a bicycle for each computer.

For Chasquack, 2 computers is greater than its opportunity cost of producing one bike (1/2); for Innduck 1/2 bike is greater than its opportunity cost of one computer (1/3).

37. B, unemployment increases; bond prices decrease. A tight monetary policy results in higher interest rates, which will reduce aggregate demand and raise the unemployment rate. When interest rates increase, bond prices decrease.

38. D, Price level decrease; Real GDP decrease. If there is a decrease in gross investment, then aggregate demand will shift left. The decrease in AD will cause a decrease in both the price level and real GDP.

39. E, Institutional money market funds. Institutional money market funds are part of the M3 definition of money along with large time deposits (and M1 + M2).

40. B, Dollar appreciates; Investment decreases. Tight monetary and expansionary fiscal policies will lead to higher interest rates, which will appreciate the international value of the dollar (foreigners demand more US interest-bearing assets). However, the higher interest rate will discourage investment spending (Ig) in the US.

41. D, II and IV. With the given information, only II and IV can be known for certain. When the GDP price index or GDP deflator is 100, then that year is the base year. In the base year, nominal GDP is equal to real GDP.

42. C, education and training. Factors that bring about growth include increases in economic resources, productivity, technology, education, and whatever fosters spending on capital goods (subsidies or tax credits that increase the level of investment).

43. C, Increased income in the eurozone. If income increases in the eurozone, then Europeans would demand more US goods and US dollars. The graph shows an increased supply of US dollars ($ depreciates) due to higher interest rates in Europe or greater demand for European goods. The expansionary monetary policy and contractionary fiscal policy indicate that there will be lower interest rates in the US.

44. D, The euro appreciated in value relative to the US dollar. The graph shows a US dollar that depreciates against the euro. The quantity of US dollars demanded increased (point-to-point movement) and the demand for euros most likely increased. Now, US dollars are cheaper for Europeans to purchase.

45. B, 2%. Use the Fisher Equation: Real Interest Rate = Nominal Interest Rate – Inflation Rate (R.I.R. = 9 - 7).

46. C, $9,000. Because the bank must hold on to $100.00 of the $1,000.00 deposit, you calculate the change in money supply by multiplying excess reserves ($900) by the money multiplier (1/0.1 = 10). The change in money supply (loans through excess reserves) is $9,000.

47. A, Disposable Income increases; Price Level increases; Output increases. This tax cut represents an expansionary fiscal policy. It will increase the disposable income of households, increase consumption, shift aggregate demand to the right, increase the price level, increase real GDP (output), and reduce unemployment.

48. B, an increase in personal income taxes. An increase in taxes means that the government needs to borrow less. There is a decrease in demand for loanable funds (or rightward shift of the private supply of loanable funds), which causes interest rates to decrease.

49. D, discount rate. The discount rate is the interest rate that the Fed charges banks for short-term loans. The Fed has the power to change this rate directly. Remember, when it's easier for banks to borrow money, it's easier for banks to lend money.

50. B, Rightward shift of the production possibilities curve. A rightward shift of the PPC and LRAS are the two key ways to show long-term economic growth.

51. B, Money stock decreased. The equation of exchange is MV = PQ (Money Stock x Velocity = Nominal GDP). If nominal GDP stayed the same when velocity increased, then money stock must have decreased.

52. B, credit; capital. A US treasury bond is a financial asset and therefore a capital account transaction (although the US records it in the financial account). Because foreign money is coming into the US, it is a credit to the capital account.

53. D, lower interest rates. Demand-pull inflation occurs when aggregate demand increases. The only choice that will cause a rightward shift of AD is lower interest rates because of increased private spending (C + Ig).

54. B, A tight US monetary policy. When the Federal Reserve pursues a tight monetary policy, interest rates rise in the US. Higher interest rates are an incentive for foreigners to purchase interest-bearing assets (treasury bonds) in the US. This means greater demand for the dollar. The USD appreciates and the Japanese yen depreciates.

55. C, Unemployment. If the federal government increases taxes or cuts spending, the government will borrow less money. This causes interest rates to fall in the loanable funds market and makes the dollar less attractive to foreigners. The contractionary fiscal policy will cause aggregate demand and the price level to fall, and unemployment to rise.

56. A, Increase by $250 million. If there is an equal increase in government spending and taxes, the government is operating with a balanced budget. The multiplier in this case is equal to 1. Remember, the spending multiplier is stronger than the tax multiplier. Real output will still increase. Simply take the change in spending ($250 million) and multiply by 1.

57. C, rise; stays the same. Rational expectation theory suggests that when the Fed increases the money supply, people expect inflation and demand higher wages. The peoples' expectations counteract the Fed's expansionary monetary policy, which results in higher prices while output and employment stay the same.

58. A, appreciated; more; deficit. It now takes less dong to buy a taka. The dong appreciated which makes it easier for the Vietnamese to buy goods from Bangladesh (import more). This will decrease net exports causing the current account balance of Vietnam to move toward a deficit.

59. E, High rate of inflation. To fight inflation, the Fed targets a higher federal funds rate by

selling bonds. This shifts the money supply to the left causing spending to fall. When spending falls, aggregate demand will decrease causing prices to fall.

60. E, demand for the dollars increases; Net exports decrease. The supply of loanable funds will decrease causing the real interest rate to rise. The higher real interest rate will increase demand for US interest-bearing assets and demand for dollars (dollar appreciates). US goods will now look more expensive to foreigners, which will cause exports to fall and imports to rise.

Free Response Solutions

1. This economy is experiencing a **recession** because the unemployment rate is high and the inflation rate is low. Short-run aggregate supply intersects aggregate demand at a point that is to the left of the long-run aggregate supply curve.

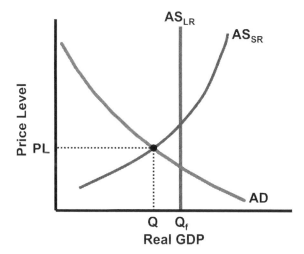

A prudent fiscal policy is to **increase government spending or lower personal income taxes.** This will increase aggregate demand, real GDP, and price level. The unemployment rate would decrease.

A sensible monetary policy is to **buy bonds on the open market,** which increases the money supply and lowers nominal interest rates. This will increase investment spending and aggregate demand.

If no policies are enacted, **nominal wages will fall in the long run.** This will **shift the short-run aggregate supply curve to the right** toward full employment.

2. Money market graph showing an increase in money demand. The rightward shift of demand will raise the nominal interest rate.

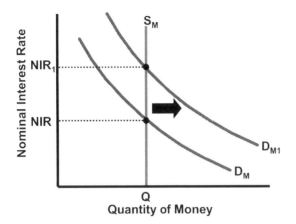

The Fed can counteract this increase in nominal interest rate by **buying bonds**. This will **shift the supply of money to the right** and keep interest rates low.

If the Fed buys $30 million worth of treasuries, the money supply will increase by **$120 million**. This is because the money multiplier is 4 (1/0.25). $30 million x 4 = $120 million.

If you deposit $200 into your checking account, the money supply will increase by **$600**. The money multiplier is 4 when the reserve ratio is 25%. Your bank will keep $50 and lend the excess reserves of $150. To get the change in money supply from a deposit, multiply excess reserves by the multiplier ($150 x 4 = $600)

3. Short-run Phillips curve shifts to the right when the short-run aggregate supply curve shifts left:

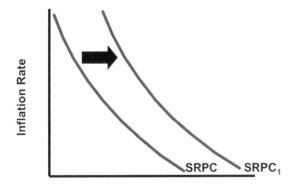

If fewer US dollars are supplied to the foreign exchange market, the value of the **dollar will appreciate**. This will make foreign goods look cheap and US goods look expensive. **Net exports would decline**.

If net exports decline, **aggregate demand will shift left**. This would lead to lower prices and more unemployment. This is **point-to-point downward movement along a short-run Phillips curve**.

No Bull Exam – Microeconomics

Part I: Answer all 60 multiple-choice questions by choosing the letter of the best answer.

1. Which of the following best describes the elasticity of a perfectly competitive firm's demand curve?
   a. Relatively inelastic
   b. Perfectly inelastic
   c. Unit elastic
   d. Perfectly elastic
   e. None of the above

2. Assume that Dorothy's marginal utility-price ratio is 7.3 for 9 bottles of water and 3.6 for 7 pairs of ballet shoes. How should she adjust her purchases of water and ballet shoes?
   a. Buy less water and more ballet shoes
   b. Buy more water and less ballet shoes
   c. Buy more water and more ballet shoes
   d. Buy less water and less ballet shoes
   e. She should not change anything

3. Suppose that perfectly competitive firms are currently producing at P1, Q1. Which of the following will occur in the long run?

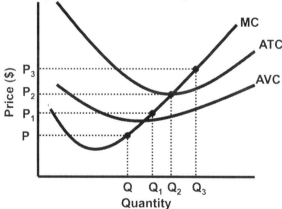

   a. Price will decrease to P
   b. Firms will enter the industry
   c. Price will increase to P3
   d. Firms will exit the industry
   e. Price will stay the same

4. Refer to the perfectly competitive firm graph in the previous question: Suppose the firm takes price P from the market. Which of the following is not true about production at P?
   a. This firm will produce Q units
   b. This firm will exit the market in the long run
   c. This firm will shut down
   d. This firm will pay fixed costs at P
   e. This firm's average variable costs exceed the market price

5. If the economy below moves from point A to point C, then how many guitars are sacrificed?

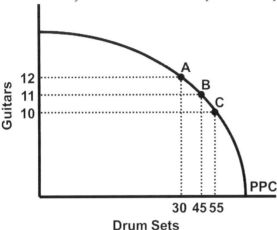

   a. 1 guitar
   b. 2 guitars
   c. 10 guitars
   d. 12 guitars
   e. 15 guitars

6. Suppose the percentage change in quantity demanded of Good AY divided by the percentage change in price of Good BE is negative. Which of the following must be true?
   a. Good A and Good B are unrelated goods
   b. Good A and Good B are substitutes
   c. Good A and Good B are complements
   d. Good A is a normal good and Good B is inferior
   e. None of the above

7. Suppose that the federal government issues a report stating that consuming carrot cake increases the chances of falling ill. What will happen to market price and quantity?
   a. Price decreases; Quantity decreases
   b. Price increases; Quantity increases
   c. Price increases; Quantity decreases
   d. Price indeterminate; Quantity decreases
   e. Price decreases; Quantity indeterminate

8. The monopolist's average revenue is
   a. less than the marginal revenue
   b. less than the price
   c. always equal to the total revenue
   d. greater than the marginal revenue
   e. greater than the price

9. Assume that the market for Good JAY is currently in equilibrium and Good JAY is a normal good. If consumer income rises at the same time that the costs of producing Good JAY falls, then what will happen to the market equilibrium?
   a. Price increases; Quantity indeterminate
   b. Price decreases; Quantity indeterminate
   c. Price indeterminate; Quantity increases
   d. Price indeterminate; Quantity decreases
   e. Price decreases; Quantity increases

10. Suppose there are many firms in a market selling similar products that are somewhat unique. Which of the following is true concerning a firm that sells its good within this market structure?
    a. Demand is above the marginal revenue curve
    b. Demand is perfectly elastic
    c. Demand is perfectly inelastic
    d. Demand is upward sloping
    e. Demand is less than the average revenue

11. Suppose that there are only two bands that record and perform rock music, The Sonic Booms and The Cranky Elves. Each band must decide whether to write songs about "Love" or "Death." The cells contain the payouts for these strategies, and the first entry in each cell represents The Sonic Booms' payout. If The Sonic Booms write about "Death" and the bands do not cooperate, what will the payout be for The Cranky Elves?

		The Cranky Elves	
		Love	Death
The Sonic Booms	Love	$105, $105	$150, $72
	Death	$72, $150	$125, $125

    a. $72
    b. $105
    c. $125
    d. $150
    e. $210

12. Refer to the game theory matrix regarding The Cranky Elves and The Sonic Booms in the previous question: What is the Nash equilibrium?
    a. Sonic Booms Love; Cranky Elves Love
    b. Sonic Booms Death; Cranky Elves Love
    c. Sonic Booms Love; Cranky Elves Death
    d. Sonic Booms Death; Cranky Elves Death
    e. A Nash equilibrium does not exist in this game

13. Suppose that there is a disease that harms tomato crops and the elasticity of demand for ketchup is less than 1. How will the market price of ketchup and total revenue from ketchup sales change due to this disease?
    a. Price constant; Total revenue constant
    b. Price decreases; Total revenue decreases
    c. Price increases; Total revenue decreases
    d. Price increases; Total revenue constant
    e. Price increases; Total revenue increases

14. The graph below shows the market for greasy bacon cheeseburgers after a per-unit tax. Which shaded region represents the consumer surplus after the tax?

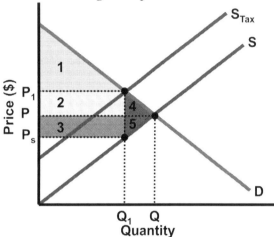

    a. 1
    b. 2
    c. 2 & 4
    d. 2 & 3
    e. 1, 2, & 4

15. Refer to the graph for greasy bacon cheeseburgers in the previous question: Which shaded region represents the consumer surplus before the tax?
    a. 1
    b. 2
    c. 2 & 4
    d. 3 & 5
    e. 1, 2, & 4

16. If the supply of good BEE and the demand for good BEE increase simultaneously, then what will happen to market price and quantity?
    a. Price decreases; Quantity indeterminate
    b. Price indeterminate; Quantity decreases
    c. Price decreases; Quantity increases
    d. Price increases; Quantity indeterminate
    e. Price indeterminate; Quantity increases

17. The additional revenue that a firm receives from hiring one more worker is known as the
    a. average variable cost
    b. marginal cost
    c. marginal revenue product
    d. marginal physical product
    e. average fixed cost

18. The chart below contains the short-run total costs for an avocado farmer that sells its output in a perfectly competitive market. If the current market price is $22, which of the following is true in the long run?

Output	Total Cost
0	45
1	51
2	64
3	78
4	98
5	120
6	147
7	175

    a. Firms will enter the market due to economic profits
    b. Firms will exit the market due to economic losses
    c. Firms will incur economic losses
    d. Firms will earn economic profits
    e. Firms will not break even

19. All of the following are examples of economic resources except
    a. a nuclear power plant in Connecticut
    b. a robotic arm in a Japanese car factory
    c. farmland in Iowa
    d. securities on a stock exchange
    e. badger hair used in the production of shaving brushes

20. Suppose that the market for digital video cameras is in equilibrium at a market price of $149. The government decides to establish a ceiling price of $175 to support low-income consumers. Which of the following will occur?
    a. The demand curve will shift right
    b. There will be a shortage of cameras
    c. There will be a surplus of cameras
    d. Underground markets will develop
    e. The market will remain at its equilibrium price

21. When a monopolist's marginal revenue is positive then
    a. demand for the good is unit elastic
    b. demand for the good is elastic
    c. demand for the good is inelastic
    d. demand for the good is perfectly inelastic
    e. none of the above

22. Average variable costs (AVC) are decreasing whenever
    a. average total costs are decreasing
    b. average total costs equal AVC
    c. marginal costs exceed AVC
    d. average total costs are increasing
    e. average total costs are to the right of marginal costs

23. Suppose that consumers are willing and able to buy 5,000 wooden baseball bats when the market price is $20, and 4,000 wooden baseball bats at a price of $30. Which of the following is true concerning the demand for wooden baseball bats over this price range?
    a. Elasticity of demand is unit elastic
    b. Elasticity of demand is price elastic
    c. Elasticity of demand is 0
    d. Elasticity of demand is price inelastic
    e. Elasticity of demand is indeterminate

24. Which of the following simultaneous shifts will cause the market price to be indeterminate and the equilibrium quantity to increase?
    a. Demand increases; Supply increases
    b. Demand decreases; Supply decreases
    c. Demand decreases; Supply increases
    d. Demand increases; Supply decreases
    e. Demand increase; Supply stay constant

25. Which of the following is not an economic function of the government?
    a. Enact laws that promote competition
    b. Maintain growth of the money supply
    c. Correct externalities
    d. Provide public goods
    e. Redistribute income toward equality

26. Suppose that Petro carefully plans his yearly gym schedule. Petro determines that he can complete 260 upper body workouts in one year or 780 lower body workouts in one year. What is his opportunity cost of completing one lower body workout?
    a. 1/3 of an upper body workout
    b. 1/2 of an upper body workout
    c. 1 upper body workout
    d. 2 upper body workouts
    e. 3 upper body workouts

27. The central government chooses to assist consumers with low income that are in the market for fuel-efficient automobiles. Which of the following courses of action should the central government take to meet its objective?
    a. Establish a price ceiling below equilibrium price
    b. Establish a price floor below equilibrium price
    c. Impose an excise tax on the sale of fuel-efficient car
    d. Impose a value-added tax on the stages of fuel-efficient car production
    e. Decrease subsidies to the producers of fuel-efficient cars

28. Refer to the information given in the previous question: What is the likely effect of the government's course of action on the market for fuel-efficient automobiles?
    a. The supply of fuel-efficient cars will shift to the right
    b. The demand for fuel-efficient cars will increase
    c. There will be a surplus of fuel-efficient cars
    d. There will be a shortage of fuel-efficient cars
    e. Nothing will happen, the market will remain in its original equilibrium

29. Which of the following is true of total revenue (TR) for a perfectly competitive rice producer?
    a. TR decreases at an increasing rate.
    b. TR decreases at a decreasing rate.
    c. TR increases at an increasing rate
    d. TR increases at a decreasing rate
    e. TR increases at a constant rate

30. Assume that the graph below depicts the prices, costs, and revenues for a company that is the only seller of glockenspiels. Which of the following is true about production at Q2?

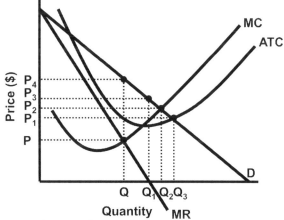

    a. Demand is inelastic
    b. Total revenue is maximized
    c. The firm will break even
    d. Demand is elastic
    e. Profits are maximized

177

31. Refer to the graph for glockenspiels in the previous question: If this firm produces at its profit-maximizing level of output, its per-unit economic profit is
    a. P4 – P
    b. P2 – P1
    c. P4 – P1
    d. P1 – P
    e. P4 – P2

32. According to the cost curves below, average variable costs

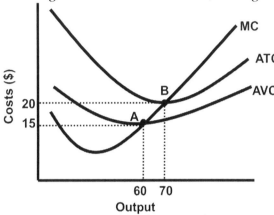

    a. intersect marginal costs at $20
    b. always exceed marginal costs
    c. increase after 60 units of output
    d. are $20 at 70 units of output
    e. exceed average total costs

33. Assume that the supply of lederhosen and the demand for lederhosen increase simultaneously. All of the following could have caused one of these shifts to occur except
    a. an improvement in productivity
    b. an increase in price of a complement
    c. an increase in number of sellers
    d. an increase in price of a substitute good
    e. an increase in subsidies to lederhosen producers

34. Suppose a nation's production possibilities curve is a straight line. Which of the following must be true?
    a. Opportunity costs are increasing
    b. Opportunity costs are decreasing
    c. Opportunity costs are constant
    d. Opportunity costs can be increasing or decreasing
    e. The economy is fully employed

35. Suppose that the government subsidizes producers of snowboards and the elasticity of demand for snowboards is greater than 1. How will the market price of snowboards and total revenue from snowboard sales change?
    a. Price increases; Total revenue increases
    b. Price decreases; Total revenue increases
    c. Price decreases; Total revenue decreases
    d. Price increases; Total revenue decreases
    e. Price increases; Total revenue stays the same

36. Which of the following would occur in a perfectly competitive labor market when a minimum wage is imposed?
    a. Supply shifts right
    b. Demand shifts right
    c. A shortage of workers
    d. A surplus of laborers
    e. Nothing, the market will maintain equilibrium

37. Which of the following is true regarding allocative efficiency and productive efficiency?
    I. When society is utilizing resources to produce the optimal combination of goods, then allocative efficiency can be achieved.
    II. A monopolist achieves productive efficiency whenever it produces at its profit-maximizing level of output.
    III. Productive efficiency means producing goods at the lowest possible cost.
    IV. Productive and allocative efficiency can never be achieved in a market economy.
    a. I and II
    b. I and III
    c. I and IV
    d. II and IV
    e. III and IV

38. What would most likely reduce the consumer surplus within a market?
    a. Increase in the supply curve
    b. Increase in the demand curve
    c. Simultaneous increase in the supply and demand curve
    d. Monopoly that becomes a perfectly competitive market
    e. Firms in an oligopoly that practice price discrimination

39. Assume that the cross elasticity of demand between Good VEE and Good UU is equal to -2, and the income elasticity of demand for Good UU is equal to 0.25. Which of the following must be true?
    a. VEE and UU are substitutes; UU is an inferior good
    b. VEE and UU are complements; UU is a normal good
    c. VEE and UU are complements; UU is an inferior good
    d. VEE and UU are substitutes; UU is a normal good
    e. None of the above

40. Which of the following is true concerning a monopolist and a perfectly competitive firm?
    I. A monopolist's marginal revenue curve is below the demand curve.
    II. A monopolist produces at an output level where demand is elastic, and a perfectly competitive firm produces at an output level where its demand is perfectly elastic.
    III. A perfect competitor's marginal revenue is equal to its average revenue.
    IV. A monopolist and a perfectly competitive firm will produce at a level of output where marginal revenue equals marginal cost.
    a. I and IV
    b. I and II
    c. II and IV
    d. II, III, and IV
    e. I, II, III, and IV

41. Suppose that Jeff has $60 of income to spend on hot dogs and fountain soft drinks at the baseball game. Hot dogs are $10 each and fountain soft drinks are $5 each. The marginal utility of his third hot dog is 60 and the marginal utility of his sixth soft drink is 30. Which of the following is true?
    a. He should purchase fewer dogs, more drinks
    b. He should purchase more dogs, fewer drinks
    c. Here utility is maximized within his income limit
    d. He cannot afford that combination of goods
    e. None of the above

42. Suppose that the supply curve of Good UU has an elasticity of 1.7 and the demand curve has a price elasticity of 0.8. If the government institutes a per-unit tax on Good UU then
    a. consumers will pay a greater portion of the tax
    b. producers will pay a greater portion of the tax
    c. consumers and producers share the tax equally
    d. supply will shift right
    e. demand will shift right

43. Assume that a ballpoint pen manufacturer in a perfectly competitive constant-cost industry is currently earning short-run economic profits. If the government imposes a lump-sum tax on firms that produce ballpoint pens, this firm's marginal costs will ____, economic profits will ____, and output will ____.
    a. increase; decrease; decrease
    b. decrease; increase; increase
    c. stay constant; decrease; stay constant
    d. stay constant; increase; stay constant
    e. increase; decrease; increase

44. The average total cost (ATC) curve is increasing
    a. when average variable costs are decreasing
    b. when marginal costs are less than ATC
    c. when average variable costs exceed ATC
    d. after it intersects the marginal cost curve
    e. after it intersects the average variable cost curve

45. When a market is monopolized, or becomes less competitive, as a result of a merger, there may be an increase in
    a. economies of scale due to lower average costs
    b. constant returns to scale due to increasing average costs
    c. diseconomies of scale due to lower average costs
    d. consumer surplus caused by a decrease in supply
    e. output due to less competition

46. Suppose Andrew can bake cupcakes and muffins. Based on the production table below, what is Andrew's opportunity cost of increasing cupcake production from 5 units to 6 units?

Cupcakes	Muffins
0	40
3	34
5	27
6	22
9	7
11	0

    a. 1 muffin
    b. 5 muffins
    c. 6 muffins
    d. 22 muffins
    e. 27 muffins

47. Assume that the market for good TEE is currently in equilibrium and good TEE is an inferior good. If consumer income rises at the same time that the costs of producing good TEE falls, then what will happen to the market equilibrium?
    a. Price decreases; Quantity indeterminate
    b. Price indeterminate; Quantity increases
    c. Price decreases; Quantity increases
    d. Price indeterminate; Quantity decreases
    e. Price increases; Quantity indeterminate

48. The chart below contains the short-run total costs for a celery farmer that sells its output in a perfectly competitive market. What is this firm's marginal cost of producing 6 units of output?

Output	Total Cost
0	45
1	51
2	64
3	78
4	98
5	120
6	147
7	175

   a. $17
   b. $20
   c. $22
   d. $27
   e. $28

49. Which of the following is a key factor that helps determine the price elasticity of supply?
   a. Substitute availability
   b. Percentage of income
   c. Necessity of the good
   d. Excess capacity
   e. All of the above

50. When total product increases at an increasing rate
   a. marginal product is decreasing
   b. average product exceeds marginal product
   c. average product is decreasing
   d. marginal product is increasing
   e. none of the above

51. When the long-run average total costs of a firm decrease as output increases, it is experiencing
   a. economies of scale
   b. diseconomies to scale
   c. constant returns to scale
   d. diminishing utility to scale
   e. the substitution effect

52. Which of the following is true of the marginal utility curve?
    a. It intersects total utility at total utility's maximum
    b. It increases, levels off, then decreases
    c. It decreases, reaches its lowest point, then rises
    d. It increases when total utility decreases
    e. It intersects average total costs at ATC's minimum

53. When marginal product exceeds average product, average product _____. If average product exceeds marginal product, then average product _____.
    a. rises; rises
    b. rises; falls
    c. falls; rises
    d. falls; falls
    e. remains the same; falls

54. Which of the following is not a characteristic of capitalism?
    a. Central planning
    b. Private ownership of resources
    c. Self-interest
    d. Invisible hand
    e. Markets and prices

55. Suppose that the market for yogurt is perfectly competitive and firms are experiencing short-run economic losses. If there is a decrease in the number of yogurt sellers in the long run, how will the marginal revenue and output for a typical seller change?
    a. MR shifts up; Output increases
    b. MR shifts up; Output decreases
    c. MR shifts up; Output remains constant
    d. MR shifts down; Output decreases
    e. MR shifts down; Output increases

56. If a firm is producing goods at the lowest costs possible then it is achieving
    a. absolute efficiency
    b. productive efficiency
    c. comparative efficiency
    d. allocative efficiency
    e. diminishing marginal utility

57. Assume that the price of digital MP3 players increases, but total revenue remains constant. Which of the following is true concerning the demand for MP3 players?
    a. Elasticity of demand is price inelastic
    b. Elasticity of demand is unit elastic
    c. Elasticity of demand is equal to infinity
    d. Elasticity of demand is price elastic
    e. Elasticity of demand is undefined

58. Assume that good OH and good DEE are produced using the same economic resources, and good DEE and good KEW are complementary goods. If the price of good KEW falls, then which of the following will occur in the market for good OH?
    a. Price increases; Quantity increases
    b. Price decreases; Quantity increases
    c. Price decreases; Quantity indeterminate
    d. Supply decreases; Price increases
    e. Demand decreases; Quantity decreases

59. Assume that microwavable meals are inferior goods and consumer income declines. Which of the following is true about the market for microwavable meals?
    a. Market price will fall
    b. Demand will shift right
    c. Supply will shift left
    d. Quantity supplied will decrease
    e. Quantity is indeterminate

60. Which of the following is not true regarding the production possibilities curves below?

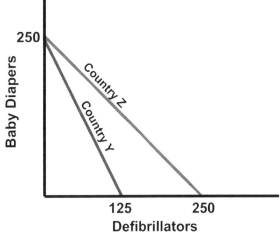

I. Resources are fixed in Country Y and Country Z
II. Opportunity costs are increasing in Country Z
III. Country Z has the absolute advantage in defibrillators
IV. An acceptable terms of trade are 1 baby diaper for 3 defibrillators
    a. I and II
    b. II and III
    c. II and IV
    d. I, II, and IV
    e. I, II, III, and IV

Part II: Answer all 3 free-response questions and include all necessary graphs and lines of economic reasoning.

1. Draw a graph of an unregulated monopoly that shows the firm taking economic losses.

a. Label the price "P" and level of output "Q."
b. Shade in the area that represents economic loss.
c. If the government wants the monopolist to produce more units of output, should the government issue lump-sum subsidies or per-unit subsidies to the producer. Explain.
d. Based on your graph in part a, what will happen to the price and quantity if the monopolist is regulated to produce at the socially optimum level of output? Explain.

2. Suppose that government researchers determine the production of a good causes harm to the environment.
   a. Draw a graph that illustrates a negative externality. Label the market price "P" and quantity "Q."
   b. Shade in the area of deadweight loss.
   c. How can the government attempt to correct this externality?
   d. Now, suppose the government was wrong. The production of the good is really great for the environment. Draw a new graph that shows a positive externality. Shade in the area of deadweight loss.

3. A perfectly competitive firm hires labor from a perfectly competitive labor market. Draw the market and the firm side by side.
   a. Identify the equilibrium wage and quantity of workers hired.
   b. If education improves the productivity of workers in the market, what will happen to the equilibrium wage and quantity of workers hired? Explain.

_____
_____
_____
_____
_____
_____
_____
_____
_____
_____
_____
_____
_____
_____
_____
_____

No Bull Exam – Microeconomics Answers

1. D, Perfectly elastic. Because a PC firm is a "price taker," the price that the firm sells its product for is constant and equal to the marginal revenue. This is also the firm's demand curve, which is perfectly elastic (horizontal). Price = Marginal Revenue = Demand for a PC firm.

2. B, Buy more water and less ballet shoes. In order for the MU/P of water to decrease from 7.3, Dorothy must purchase more water. This is because of diminishing marginal utility. In order for the MU/P of ballet shoes to increase from 3.6, she must purchase fewer shoes. The ratios will eventually become equal.

3. D, Firms will exit the industry. If PC firms incur economic losses in the short run, firms will exit the industry in the long run. The market supply will shift left raising the price to P2, bringing the firms that are left to long-run equilibrium.

4. A, This firm will produce Q units. The marginal cost curve above the minimum AVC makes up the firm's short-run supply curve. Anything below minimum AVC indicates no production and the firm will shut down. The firm will still pay fixed costs and exit the industry in the long run.

5. B, 2 guitars. 12 guitars (at point A) minus 10 guitars (at Point C) = 2 guitars.

6. C, Good A and Good B are complements. The cross elasticity of demand formula that is stated in this question is a way to determine if goods are substitutes or complements. If the cross elasticity of demand is negative, the goods are complements.

7. A, Price decreases; Quantity decreases. If rational consumers were to learn that a product is harmful to one's health, then demand for that product would decrease. Consumer tastes and preferences would cause demand to shift left (Price decreases; Quantity decreases).

8. D, greater than the marginal revenue. The average revenue curve is the demand curve and the monopolist's price is on this curve. Its marginal revenue curve is below the demand curve.

9. C, Price indeterminate; Quantity increases. The demand and supply curves will shift right. Price will be indeterminate (increase, decrease, or stay the same) and quantity will increase.

10. A, Demand is above the marginal revenue curve. Monopolistic competition is a market structure where many firms (less than the number in perfect competition) sell similar products that are not identical to one another. The marginal revenue curve is below the demand curve, which should be relatively elastic due to the presence of substitutes.

11. D, $150. The Cranky Elves will play their dominant strategy, which is to write about "Love." Therefore, the band's profit will be $150.

12. A, Sonic Booms Love; Cranky Elves Love. The dominant strategy is the one with the highest payout regardless of the other player's strategy. If The Sonic Booms write about "Love," profits will be $105 or $150, which is greater than $72 and $125 respectively if they

wrote about "Death." The Cranky Elves face the same payouts. If each group plays "Love," they are playing their dominant strategies and that cell represents the Nash equilibrium.

13. E, Price increases; Total revenue increases. The disease to tomato crops will shift the supply of ketchup to the left. The price of ketchup will increase. Since demand for ketchup is price inelastic (Ed < 1), total revenue will increase when the price rises.

14. A, 1. The triangle marked with 1 represents the consumer surplus after the tax. It is the area under the demand curve and above the market price with the tax.

15. E, 1, 2, & 4. Regions 1, 2, and 4 represent the original consumer surplus prior to the tax. This is the area under the original demand curve and above the original market equilibrium. After the tax is enacted, the consumer surplus shrinks to region 1. This leads to deadweight loss (assuming there are no negative externalities), which is shown in regions 4 and 5.

16. E, Price indeterminate; Quantity increases. If supply shifts right, the price decreases and quantity increases. If demand shifts right, the price increases and quantity increases. Therefore, price is indeterminate because it can increase, decrease, or remain constant. The market quantity will definitely increase.

17. C, marginal revenue product. The MRP is the change in total revenue from employing one more economic input (such as a laborer).

18. B, Firms will exit the market due to economic losses. If the market price is $22 then this firm will produce 5 units. Its total revenue for 5 units is $110 ($22 x 5). Total costs are $120 at 5 units so this firm is incurring an economic loss of $10. In the long run, the inefficient firms taking economic losses will exit the industry and the market price will increase. The remaining firms will break even.

19. D, securities on a stock exchange. Financial assets and money are not considered economic resources. Capital consists of tools, machines, and factories.

20. E, The market will remain at its equilibrium price. In order for a ceiling to be effective or binding, it must be less than the equilibrium price. If a ceiling is set above the free market price, the market equilibrium stays the same.

21. B, demand for the good is elastic. When marginal revenue is positive, demand is price elastic. The unregulated monopolist will always produce along the elastic region of the demand (average revenue) curve.

22. A, average total costs are decreasing. Average total costs are decreasing when average variable costs are decreasing. Average total costs are greater than AVC at this point, and the AVC will intersect the marginal cost curve before the ATC does. It is always a good idea to sketch the graphs when answering questions like this.

23. D, Elasticity of demand is price inelastic. In this example, consumers are relatively unresponsive to changes in the price of baseball bats (elasticity is less than 1). At a price of $20, total revenue is $100,000 ($20 x 5,000). When the price increases to $30, total revenue increases to $120,000 ($30 x 4,000). Because total revenue rises when price rises, demand is

inelastic.

24. A, Demand increases; Supply increases. If demand and supply both increase (shift right), then quantity will increase. However, price can increase from a rightward shift of demand, or decrease from the rightward shift of supply. Or, price can stay the same. Therefore, we say that price is indeterminate.

25. B, Maintain growth of the money supply. The money supply is set through monetary policy of the Federal Reserve. In economics, monetary policy is different from fiscal policy. Fiscal policy consists of actions by the government to boost aggregate demand (taxes and government spending).

26. A, 1/3 of an upper body workout. Because you are given a problem consisting of total output, the quantity of lower body workouts goes in the denominator since you want to find the opportunity cost of one lower body workout (260 upper body workouts divided by 780 lower body workouts = 1/3 of an upper body workout).

27. A, Establish a price ceiling below equilibrium price. An effective price control for this situation is a binding price ceiling, which must be set below the free market equilibrium. Fuel-efficient car prices will be artificially low and result in a shortage of fuel-efficient cars.

28. D, There will be a shortage of fuel-efficient cars. Effective price ceilings below the free market equilibrium price will result in a shortage.

29. E, TR increases at a constant rate. PC firms are "price takers." The market tells the firm the price it will sell their goods at. Marginal revenue increases by the same amount as the price (P = MR). Therefore, total revenue (P x Q) increases at a constant rate.

30. A, Demand is inelastic. At Q2, marginal revenue is negative. When MR is negative, the demand curve is relatively price inelastic. An unregulated monopolist will only produce where MR is positive, which is in the elastic region of the demand curve.

31. C, P4 - P1. An unregulated monopolist will produce at MR = MC with the price located on the demand curve (P4, Q). Profit is total revenue minus total cost. P4 represents revenue and P1 represents cost. Therefore, P4 - P1 tells us the per-unit economic profit. If you multiply that number by Q then you would get the total economic profit.

32. C, increase after 60 units of output. The AVC curve meets the marginal cost curve at 60 units of output, which is the minimum of the AVC curve. After 60 units of output, AVC must increase.

33. B, an increase in price of a complement. If the price of a complement good rises, then the demand for lederhosen would decrease.

34. C, Opportunity costs are constant. The law of increasing opportunity costs applies to PPCs that are bowed outward from the origin. Straight-line production possibility frontiers indicate an economy with constant opportunity costs.

35. B, Price decreases; Total revenue increases. If the government subsidizes the sellers of

snowboards, supply will shift right and lower the market price. Because elasticity of demand is greater than 1 (price elastic), total revenue will increase when the price falls. Consumers are relatively responsive to the price change.

36. D, A surplus of laborers. A minimum wage is a price floor (legal minimum price above equilibrium) so there will be a surplus of laborers in the labor market.

37. B, I and III. Allocative efficiency is producing the optimal combination of goods and services for society (P = MC). Productive efficiency is producing at lowest cost (P = Minimum ATC).

38. E, Firms in an oligopoly that practice price discrimination. When a firm with market power price discriminates, it charges buyers different prices for the same product. This will eat up the consumer surplus and increase producer surplus.

39. B, VEE and UU are complements; UU is a normal good. If the cross-price elasticity of demand is negative, then the two goods are complements. When the income elasticity of demand is positive, the good is a normal good.

40. E, I, II, III, and IV. All of the statements are true. The average revenue is equal to total revenue divided by the quantity, which makes up the price or demand curve.

41. C, Here utility is maximized within his income limit. The MU/P of hotdogs is 6 (60/10) at 3 units and the MU/P of soft drinks is 6 (30/5) at 6 units. This is ideal because you want the MU/P ratios for hotdogs and soft drinks to be equal. Additionally, this is within Jeff's income constraint of $60 ($30 for 3 hot dogs + $30 for 6 soft drinks).

42. A, consumers will pay a greater portion of the tax. When demand is inelastic (less than 1) and supply is elastic (greater than 1), the consumer pays a larger portion of the tax than the seller because consumers are not as responsive to changes in price.

43. C, stay constant; decrease; stay constant. Because the tax is a lump-sum tax, marginal costs are unaffected by the tax, but average total costs will shift up. The firm will produce the same output but its short-run profits will decrease.

44. D, after it intersects the marginal cost curve. After the average total cost curve intersects the marginal cost curve, ATC is increasing and marginal costs exceed ATC.

45. A, economies of scale due to lower average costs. When a firm becomes so large that it controls an entire market, the long-run average total costs can decrease. The firm is enjoying economies of scale, which makes it very difficult for smaller firms to compete. It serves as a barrier to entry. Consumer surplus will decrease, output will most likely decrease, and market price will increase.

46. B, 5 muffins. 27 muffins minus 22 muffins = 5 muffins.

47. A, Price decreases; Quantity indeterminate. When the costs of production fall, supply shifts to the right (P falls, Q rises). Since TEE is an inferior good, demand shifts left when income increases (P falls, Q falls). After both shifts have occurred, the quantity can increase,

decrease, or stay the same.

48. D, $27. The marginal cost of the sixth unit equals the change in total cost divided by change in output. Change in total cost is $27 (147-120) and change in output is 1 (6 - 5). 27 divided by 1 equals $27.

49. D, Excess capacity. Excess capacity means that firms can produce more goods; some resources are currently not being used to full potential. The other choices are determinants of the price elasticity of demand.

50. D, marginal product is increasing. When total product increases at an increasing rate, the marginal product increases. Once the marginal product curve reaches its max, it will level off then begin to fall. The total product curve will continue to increase, but at a decreasing rate (diminishing marginal returns).

51. A, economies of scale. In the long run, a firm can experience economies of scale when ATC decreases as output increases. Monopolies often enjoy economies of scale because they can produce at such a large scale efficiently. It also serves as a barrier to entry in markets that are controlled by one or a few sellers.

52. B, It increases, levels off, then decreases. The marginal utility increases then reaches a maximum, levels off, then decreases. It will eventually reach zero and become negative.

53. B, rises; falls. These two curves intersect where average product is at its maximum.

54. A, Central planning. Central planning is a term most often associated with command economies.

55. A, MR shifts up; Output increases. The market supply will shift left causing the price to rise. Marginal revenue equals price for a perfectly competitive firm so marginal revenue will shift up. The firm will produce more as it moves along its marginal cost curve to a new output level. This will ultimately be the long-run equilibrium.

56. B, productive efficiency. Productive efficiency means producing at lowest cost possible (the price will equal minimum average total costs).

57. B, Elasticity of demand is unit elastic. When prices change and total revenue remains constant, then the price elasticity of demand is unit elastic over that price range.

58. D, Supply decreases; Price increases. Since the price of good KEW falls, demand for good DEE increases. There will be greater demand for resources to produce good DEE, which increases the resource costs of good OH. Therefore, the supply of OH will shift to the left, increasing the market price and decreasing quantity.

59. B, Demand will shift right. If income falls, demand for inferior goods will increase (Price increases; Quantity increases).

60. C, II and IV. The terms of trade are unacceptable to Country Z. They would receive only 1/3 of a baby diaper for each defibrillator they export. Z needs to import more than 1 baby

diaper for each life-saving device.

Free Response Solutions

1. Monopoly operating at an economic loss:

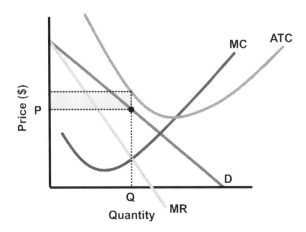

The monopoly above is operating at an economic loss because the price is less than average total costs at the level of output where MR = MC.

If the government wants the monopolist to increase its output, it should **increase per-unit subsidies to the monopolist**. This will lower marginal costs, shifting the MC curve downward. Lump-sum subsidies will not alter marginal costs.

If the monopolist is regulated to produce at the socially optimal level of output, then the monopolist would produce where price equals marginal costs. Compared to the unregulated level of output, the **price will decrease and quantity will increase**.

2. Negative Externality Graph:

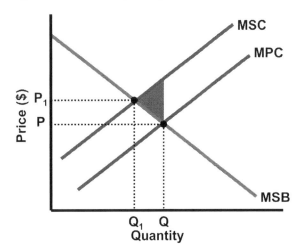

Marginal social costs exceed marginal social benefits when negative externalities exist. Deadweight loss is shown by the shaded triangle.

The government can correct this negative externality by **taxing production of the good to increase production costs**.

Positive Externality Graph:

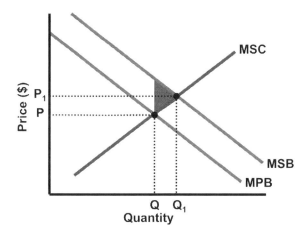

When there are positive externalities present, the **marginal social benefits exceed the marginal social costs**. The deadweight loss is shown as the shaded triangle.

3. Perfectly Competitive Labor Market and Firm:

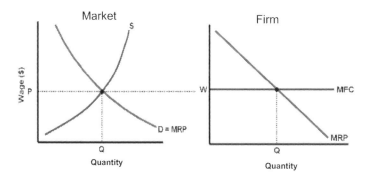

The market sets the wage, which becomes the firm's marginal factor cost (MFC) or supply curve.

If there is an increase in productivity of the workers, the demand for workers increases. This will cause the **market wage and quantity to increase**.

No Bull Review Sheet – Macroeconomics

Graphs

Production Possibilities (p.6), Market Equilibrium (p.9), AD/AS (p.33), Money Market (p.50), Loanable Funds Market (p.62), Phillips Curve (p.64), Foreign Exchange Market (p.77)

Concepts

Basic Concepts: Scarcity (p.6), Economic Resources (p.6), Opportunity Cost (p.6), Law of Increasing Costs (p.6), Absolute Advantage (p.7), Constant Cost PPC (p.7), Comparative Advantage (p.8), Specialization & Trade (p.8), Supply & Demand (p.9), Shortage (p.10), Surplus (p.10), Determinants of Demand (p.11), Determinants of Supply (p.12), Price Ceiling (p.14), Price Floor (p.15)

Econ Performance: Circular Flow Model (p.23), Factor Market (p.23), Product Market (p.23), Gross Domestic Product (p.23), Gross Investment (p.24), Business Cycle (p.24), Frictional Unemployment (p.25), Structural Unemployment (p.25), Cyclical Unemployment (p.25), Consumer Price Index (p.25), Demand-Pull Inflation (p.25), Cost-Push Inflation (p.25), Stagflation (p.25), Effects of Unanticipated Inflation (p.26)

AD/AS & Fiscal Policy: Short-Run Aggregate Supply (p.33), Long-Run Aggregate Supply (p.33), Recession Short Run & Long Run (p.34), Inflation Short Run & Long Run (p.35), Determinants of Aggregate Demand (p.36), Determinants of Aggregate Supply (p.37), Expansionary Fiscal Policy (p.37), Contractionary Fiscal Policy (p.38), Marginal Propensity to Consume & Save (p.38), Multiplier Effect (p.38)

Banking & Monetary Policy: Money Supply (p.49), Balance Sheet (p.49), Demand Deposits (p.49), Excess Reserves (p.49), Assets (p.50), Liabilities (p.50), Bond Prices & Interest Rates (p.50), Open Market Operations (p.51), Federal Funds Rate (p.51), Discount Rate (p.51), Reserve Requirement (p.51), Expansionary Monetary Policy (p.52), Contractionary Monetary Policy (p.52)

Policies & Growth: Crowding Out Effect (p.62), Classical Economic Theory (p.65), Monetary Rule (p.66), Rational Expectations (p.66), Quantity Theory of Money (p.66), Supply-Side Economics (p.66), Economic Growth (p.66), Capital Stock (p.66)

International Sector: Balance of Payments (p.76), Current Account (p.76), Capital Account (p.76), Factors Leading to Appreciation (p.77), Factors Leading to Depreciation (p.78)

Formulas

Nominal GDP (p.23), Real GDP (p.24), Unemployment Rate (p.25), Inflation Rate (p.25), Fisher Equation (p.26), Spending Multiplier (p.38), Tax Multiplier (p.38), Change in Output (p.39), Balanced Budget Multiplier (p.39), Money Multiplier (p.50), Equation of Exchange (p.65)

No Bull Review Sheet – Microeconomics

Thank you for choosing **No Bull Review** for your studying and test prep needs. We hope you found the No Bull Approach helpful and effective.

Good Luck!

Made in the USA
San Bernardino, CA
22 February 2019